D. A Saunders

Ferns and Flowering Plants of South Dakota

D. A Saunders

Ferns and Flowering Plants of South Dakota

ISBN/EAN: 9783337002336

Printed in Europe, USA, Canada, Australia, Japan

Cover: Foto ©berggeist007 / pixelio.de

More available books at **www.hansebooks.com**

(So. Dak. Bul. No 64.)

April, 1899. Bulletin 64.

U. S.
EXPERIMENT STATION
SOUTH DAKOTA.

IN CONNECTION WITH THE
SOUTH DAKOTA AGRICULTURAL COLLEGE.

FERNS AND FLOWERING PLANTS OF SOUTH DAKOTA.

DEPARTMENT OF BOTANY AND ENTOMOLOGY.

BROOKINGS, SOUTH DAKOTA.

SIOUX FALLS, S. D.
WILL A. BEACH, PRINTER AND BINDER.
1899.

GOVERNING BOARD.

REGENTS OF EDUCATION.

Hon. H. H. Blair, Pres .... Elk Point
Hon. M. F. Greeley............................. Gary
Hon. R. W. Haire, Sec'y............. .... Aberdeen
Hon. L. M. Hough .... Sturgis
Hon. F. A. Spafford....................... Flandreau

STATION COUNCIL.

Dr. F. A. Spafford, Regent Member.
Jno. W. Heston, President of College.

Jas. H. Shepard, Director .... Chemist
E. C. Chilcott, Vice Director .... Agriculturist
D. A. Saunders.............. Botanist and Entomologist
E. L. Moore................................ Zoologist
N. E. Hansen .... Horticulturist
E. A. Burnett..................... Animal Husbandry

A. M. Allen, Secretary and Accountant.

ASSISTANTS.

A. B. Holm .... Soils
W. H. Knox................................ Chemistry
S. A. Cochrane .... Irrigation
W. S. Thornber .... Horticulture
F. G. Orr.................................. Librarian

☞ Any farmer of the state can have the Bulletins of this Station free upon application to the Director.

FERNS AND FLOWERING PLANTS OF SOUTH DAKOTA.

DEPARTMENT OF BOTANY AND ENTOMOLOGY.

D. A. SAUNDERS.

INTRODUCTION.

The results embodied in the accompanying catalogue are based on specimens in the College herbarium, and on notes taken in the field by Professor T. A. Williams and the writer.

The nomenclature followed is the modification of the Paris Code adopted by the Botanical Club of the American Association for the Advancement of Science at Rochester in 1892 and amended at the Madison, Wis., meeting in 1893. The only synonyms given are such as would be a help to one accustomed to using Gray's Manual.

In preparing this catalogue, Dr. P. A. Rydberg's Flora of the Black Hills* has been freely used. Specimens have also been received from various collectors in that region. Professor Williams has made extensive collections in the Sioux valley, has done considerable work along Bigstone Lake, and has made one trip from Pierre to Rapid City and return, obtaining many interesting specimens, especially from the Bad Lands. Messrs. David Griffiths, Earl Douglass, Jno. J. Thornber, students of Professor Williams, have done considerable collecting east of the Missouri river. Mr. L. W. Carter has made various collecting trips in the eastern part of the state, and, in company with Mr. Griffiths, one extended trip from Forest City along the Moreau river to the Black Hills, returning along the Cheyenne river to Fort Pierre. The

---
*Rydberg, P. A., Contributions from the United States National Herberium, Vol. III., No. 8.

writer has collected over most of the state east of the Missouri river, and, in company with Mr. J. R. Towne, in the summer of 1897 made very careful examinations of the Little Minnesota river and its tributaries. On this trip many eastern species heretofore unknown to the state were obtained.

Acknowledgment is due Dr. N. L. Britton, Dr. P. A. Rydberg, and Messrs. Nash and Bicknell of the New York Botanical Garden, Dr. L. H. Bailey of Cornell University, and Professor Sargent of the Arnold Arboretum for the verification and determination of certain species.

CATALOGUE.

PTERIDOPHYTES. — THE FERNS AND THEIR ALLIES.

OPHIOGLOSSACEÆ.—The Adder-Tongue Family.

Botrychium matricarifolium, A., Br., Matricary Grape-fern.
Two specimens doubtfully referred to this species were collected near Custer, in the Black Hills, by Dr. Rydburg.

Botrychium virginianum, (L.) Sw. Virginian Grape-fern.
In wooded ravines in the Minnesota region and in the Black Hills; rare.

POLYPODIACEÆ.—The True Ferns.

Onoclea sensibilis, L., Sensitive fern.
Near Rapid City, in the Black Hills.

Onoclea struthiopteris, (L.) Hoffm. Ostrich-fern.
Along shaded streams in the Minnesota Valley and the Black Hills.

Woodsia scopulina, D. C. Eaton. Rocky Mountain Woodsia.
On wooded hillsides in the Black Hills; rare.

Woodsia Oregon, D. C. Eaton. Oregon Woodsia.
Common throughout the Black Hills.

Cystopteris fragilis, (L.) Bernh. Brittle-fern.
On damp shady banks bordering streams and lakes throughout the state.

Phegopteris dryopteris, (L.) Fee. Oak-fern.
In deep woods in the Black Hills.

Dryopteris Felix-Mas, (L.) Schott. Male-fern.
   Among rocks in the Black Hills.

Asplenium trichomanes, L. Maiden-hair spleanwort.
   Crevices of rocks in the Black Hills.

Asplenium Felix-foemina, (L.) Fee. Bernh. Lady-fern.
   In the Black Hills.

Asplenium Septentrionale, (L.) Hoffm. Northern spleanwort.
   In crevices of rocks in the Black Hills.

Adiantum Capillus-Veneris, L. Venus-hair fern.
   Along a warm stream, near Cascade, in the Black Hills.*

Pteris aquilina, L. Brake.
   Near Custer, in the Black Hills.

Pellaea atropurpurea, (L.) Link. Purple-stemmed Cliff-brake.
   Common in crevices of rocks in the Black Hills.

Pellaea Breweri, Eaton. Brewer's Cliff-Brake.
   In limestone crevices near Bull Spring, in the Black Hills.

Cheilanthes gracilis, (Fee.) Mett. Slender Lip-fern.
   On exposed rocks in the Black Hills.

Polypodium vulgare, L. Common Polypody.
   In crevices of rocks in the Black Hills.

Polypodium vulgare rotundatum, Wilde. Round-lobed Polypody.
   In crevices of rocks near Custer, in the Black Hills.

MARSILEACEÆ.—Marsilea Family.

Marsilea vestita, Hook & Grev. Hairy Marsilea.
   Very abundant in swails and ditches from the James river valley westward to the Black Hills.

*Bessey, C. E., Bot., Gaz., Vol. XXVI, No.3, 211.

EQUISETACEÆ.—Horse-Tail Family.

Equisetum arvense, L. Field Horsetail.
In low sandy soil in the Minnesota and Sioux valleys, and in the Black Hills.

Equisetum sylvaticum, L. Wood Horsetail.
In moist woods in the Black Hills.

Equisetum fluviatile, L. Swamp Horsetail.
In swamps in the Minnesota region.

Equisetum robustum, A., Br. Stout Scouring-rush.
In wet places throughout the state; less common than the next.

Equisetum laevigatum, A., Br. Smooth Scouring-rush.
In low wet places; very abundant throughout the state.

LYCOPODIACEÆ.—Club Moss Family.

Lycopodium obscurum, L. Ground Pine.
In moist woods in the Black Hills.

SELAGINELLACEÆ.—The Little Club Mosses.

Selaginella rupestris, (L.) Spring. Rock Selaginella.
On rocky slopes in the Black Hills.

SPERMOPHYTES.—SEED BEARING PLANTS.

GYMNOSPERMÆ.—GYMNOSPERMS.

PINACEÆ.

Pinus ponderosa scopulorum, Englm. Western Yellow pine.
Throughout the Black Hills.

Picea Canadensis, (Mill.) B. S. P. White Spruce.
In the Black Hills, especially in the northern part.

Juniperus nana, Willd. Juniper.
Juniperus Sibericus. Burgsd.
  On dry knolls in the Black Hills.

Juniperus Virginiana. Red Cedar.
  Rare in the Black Hills proper, common in the foothills and in the vicinity of streams from the Black Hills eastward to the Missouri river. It is found occasionally on the bluffs on the east side of the river.

Juniperus Sabina prostrata, (Pers.) Loud. Creeping Red Cedar.
  On dry foothills in the Black Hills.

ANGIOSPERMÆ.—THE TRUE FLOWERING PLANTS.

MONOCOTYLEDONES.—MONOCOTYLEDONS.

TYPHACEÆ.—Cat-Tail Family.

Typha latifolia, L. Broad leaved Cat-tail.
  In marshes throughout the state.

SPARGANIACEÆ.—The Burreed Family.

Sparganium eurycarpum, Englm. Broad fruited Burreed.
  In swamps, marshes and along streams throughout the state.

NAIADACEÆ. The Naiad Family.

Potamogeton lonchites, Tuckerm. Long-leaved Pondweed.
  In ponds and slow streams throught the state.

Potamogeton amplifolius, Tuckerm. Large-leaved Pondweed.
  In ponds and slow streams throughout the state.

Potamogeton heterophyllus, Schreb. Various-leaved Pondweed.
: In ponds and lakes from the Missouri river eastward.

Potamogeton perfoliatus Richardsonii, A. Bennet. Clasping-leaved Potamogeton.
: Common in the Minnesota and Sioux Valleys.

Potamogeton alpinus, Balbis. Northern Pondweed.
: In the Sioux river near Brookings.

Potamogeton foliosus, Raf. Leafy Pondweed.
: In ponds and streams throughout the state.

Potamogeton foliosus niagarensis, (Tuckerm.) Morong.
: In the Sioux Valley.

Potamogeton zosteraefolius, Schum. Ell-grass Pondweed.
: In the Sioux and James valleys.

Potamogeton Hillii, Morong. Hill's Pondweed.
: In the Sioux valley.

Potamogeton pusillus, L. Small Pondweed.
: Common in ponds from the Missouri valley eastward.

Potamogeton pectinatus, L. Fennel-leaved Pondweed.
: In lakes and streams throughout the state.

Potamogeton marinus occidentalis, Robbins. The Western Pondweed.
: In the Black Hills.

Potamogaton palustris, L. Swamp Patamogeton.
: In shallow water in the Sioux Valley.

Zanichellia palustris, L. Zanichellia.
: In brackish or fresh water ponds throughout the state.

Ruppia occidentalis, S. Wat. Western Ruppia.
: In a strongly alkaline pond in the Minnesota valley, near Gary.

Naias flexilis, (Willd.) Rost & Schmidt. Slender Naias.
: In ponds and lakes throughout the state.

SCHEUCHZERIACEÆ —Arrow-Grass Family

Triglochin palustris, L. Marse Arrow-grass.
In bogs in the Minnesota Valley.

Triglochin maritima, L. Seaside Arrow-grass.
In marshes throughout the state.

ALISMACEÆ.—Water-Plantain Family.

Alisma plantago-aquatica, L. Water-plantain.
Common in the edges of lakes and streams throughout the state.

Echinodorus cordifolius, (L.) Griseb. Upright Burhead.
Occasional in the southeastern corner of the state, Sioux Falls; Yankton.

Lophotocarpus calycinus, (Engelm.) J. G. Smith. Lopotocarpus.
Occasional in swamps from the Missouri river eastward.

Sagittaria arifolia, Nutt. Arum-leaved Arrowhead.
In the Minnesota and James valleys and in the Black Hills.

Sagittaria graminea, Michx. Grass-leaved Sagittaria.
In shallow waters from the Missouri valley eastward.

Sagittaria latifolia, Willd. Broad-leaved Arrowhead.
In shallow water throughout the state.

VALLISNERIACEÆ. Tape-Grass Family.

Philotria Canadensis, (Michx.) Britton. Waterweed, Ditchmoss.
Elodea Canadensis, Michx.
In the Minnesota and Sioux river valleys.

Vallisneria spiralis, L. Tape-grass. Ell-grass.
In Big Stone lake, in the Minnesota valley.

GRAMINEÆ.—Grass Family.

Andropogon Hallii, Hack. Hall's Beardgrass. Turkey-foot Grass.
: In the Bad Lands. Probably more or less abundant from the Missouri valley westward to the Black Hills.

Andropogon scoparius, Michx. Broom Beardgrass.
: Abundant throughout the state. A valuable forage grass.

Andropogon furcatus, Muhl. Forked Beardgrass.
: Abundant from the Missouri eastward.

Chrysopogon avenaceus, (Michx.) Benth. Bushy Bluestem, Indian Grass.
: In low damp ground from the Missouri valley eastward. Locally abundant.

Syntherisma linearis, (Krock.) Nash. Small Crab Grass.
Panicum glabrum. Gaud.
: Sparingly introduced into the southeastern part of the state. Near Yankton and Sioux Falls.

Syntherisma sanguinalis, (L.) Nash. Finger Grass; Large Crab Grass.
Panicum sanguinale, L.
: Introduced into yards from the Missouri valley eastward.

Panicum Crus-galli, L. Barnyard Grass.
: Abundant in waste places throughout the state.

Panicum Scribnerianum, Nash. Scribner's Panicum.
: Abundant on prairies in the Minnesota and Sioux valleys.

Panicum Wilcoxianum, Vasey. Wilcox's Panicum.
: On prairies in the Minnesota valley; rare.

Panicum dichotomum, L. Forked Panicum.
: In the Minnesota and Sioux valleys.

Panicum viscidum, Ell. Velvety Panicum.

Panicum Scoparium, Michx.
: In the Sioux valley and the Black Hills.

Panicum depauperatum, Muhl. Starved Panicum.
: In dry soils in the Sioux and James valleys and in the Black Hills.

Panicum virgatum, L. Tall Smooth Panicum.
: In low ground throughout the state.

Panicum Capillare, L. Witch Grass.
: A common weed in waste places throughout the state.

Ixophorus glaucus, (L.) Nash. Foxtail. Pigeon grass.
Setaria glauca, L.
: A miserable weed in cultivated grounds throughout the state.

Ixophorus Viridis, (L.) Nash. Green Foxtail.
Setaria Viridis, L.
: Same range as the last.

Ixophorus Italicus, (L.) Nash. Hungarian Grass.
Setaria Italica, R. & S.
: Escaped along roadsides in the Sioux valley.

Cenchrus tribuloides, L. Sand Burr.
: In waste and sandy places throughout the state. More abundant westward.

Zizania aquatica, L. Wild Rice. Indian Rice. Water Oats.
: In shallow water from the Missouri river eatward.

Homalocenchrus virginicus, (Willd.) Britton. White Grass.
Leersia viriginica, R. & S.
: In shallow water in the Minnesota and Sioux valleys.

Homalocenchrus oryzoides, (L.) Poll. Rice Cut Grass.
Leersia oryzoides, L.
: In swampy ground from the Missouri river eastward.

Phalaris arundinacea, L. Reed. Canary Grass.
: From the Missouri valley eastward and in the Black Hills.

Phalaris Canariensis, L. Canary Grass.
A European grass which has escaped in the eastern part of the state.

Savastana odorata, (L.) Scribn. Holy Grass. Seneca Grass. *Hierochloe borealis.* R. & S.
From the Missouri valley eastward and in the Black Hills.

Aristida purpurea, Nutt. Purple Aristida.
From the James valley westward, an abundant grass on the range. It forms large bunches of dry wiry leaves that cattle will not eat or the mower cut.

Stipa Macounii. Macoun's Stipa.
In the Black Hills.

Stipa viridula, Trin. Green Stipa.
Throughout the state.

Stipa avenacea, L. Black-oat Grass.
In the Minnesota region.

Stipa comata, Trin. & Rupr. Western Stipa.
From the Missouri valley westward.

Stipa Spartea, Trin. Porcupine Grass.
In the Minnesota, Sioux, James and Missouri valleys and in the Black Hills.

Oryzopsis micrantha, (Trin. & Rupr.) Thurb. Small flowered Mountain Rice.
From the Missouri valley westward.

Oryzopsis asperifolia, Michx. White grained Mountain Rice.
In the Black Hills.

Oryzopsis melanocarpa, Muhl. Black-fruited Mountain Rice.
On wooded bluffs in the Minnesota valley.

Oryzopsis cuspidata, (Nutt.) Vasey. Silky Oryzopsis.
Throughout the state.

Oryzopsis Juncea, (Michx.) B. S. P. Slender Mountain Rice.
In the Black Hills.

Muhlenbergia Mexicana, (L.) Trin. Meadow Muhlenbergia.
In woods and thickets in the Minnesota, Sioux and James valleys.

Muhlenbergia racemosa, (Michx.) B. S. P.
From the Missouri valley eastward and in the Black Hills.

Muhlenbergia ambigua, Torr. Minnesota Muhlenbergia.
In the Minnesota region.

Muhlenbergia Sylvatica, Torr. Wood Muhlenbergia.
In the Minnesota, Sioux and James valleys.

Muhlenbergia Wrightii, Vasey. Wright's Muhlenbergia.
In the Black Hills.

Brachyelytrum erectum, (Schred.) Beauv. Brachelytrum.
In the Minnesota and Sioux valleys and the Black Hills.

Phleum pratense, L. Timothy.
Escaped along streams and road sides in the Minnesota and Sioux valleys.

Alopecurus geniculatus, L. Marsh Fox-tail.
In the Sioux valley.

Sporobolus asper, (Michx.) Kunth. Rough Rush-grass.
Abundant in the James and Missouri river valleys; occasional throughout the state.

Sporobolus vaginæflorus, (Torr.) Vasey. Sheathed Rush-grass.
It occurs in the Sioux and the James river valleys; rare.

Sporobolus cuspidatus, Torr. Prairie Rush-grass.
In dry soils throughout the state.

Sporobolus neglectus, Nash. Small Rush-grass.
In the Minnesota region. Collected but once along an old Indian trail.

Sporobolus airoides, Torr. Hair-grass. Dropseed.
In the Missouri river to the Black Hills.

Sporobolus cryptandrus, (Torr.) Gray. Sand Dropseed.
In sandy soil from the Missouri river eastward.

Sporobolus heterolepis, Gray. Northern Dropseed.
In low prairies throughout the state from the Missouri river eastward. In the Minnesota region it often forms a large part of the lowland hay.

Sporobolus asperifolius, (Nees & Meyen.) Thurber. Rough-leaved Dropseed.
In dry soils from the James river valley westward.

Cinna Arundinacea, L. Wood Reed-grass.
In the southern part of the Sioux valley.

Cinna latifolia, (Trev.) Griseb. Slender Wood Reed-grass.
Cinna Pendula, Trin.
In woods in the Sioux valley near Brookings.

Agrostis alba, L. Red-top.
Sparingly introduced in the Minnesota and Sioux river valleys.

Agrostis exerata, Trin. Rough-leaved Bent-grass.
In the Black Hills.

Agrostis canina, L. Brown Bent-grass.
In the Missouri river valley; rare.

Agrostis perennans, (Walt.) Tuckerm. Thin-grass.
In rich woods in the Minnesota valley and the Black Hills.

Agrostis hyemalis, (Walt.) B. S. P.   Rough Hair-grass.
Agrostis scabra. Willd.
   A rather uncommon grass from the Missouri river eastward.

Calamagrostis macouniana, Vasey.   Macoun's Reed-grass.
   From the Missouri river eastward.

Calamagrostis Canadensis, (Michx.) Beauv.   Blue Joint.
   Common in low places, from the Missouri river eastward.

Calamagrostis breviseta, (Gray.) Scrib.   Pickering's Reed-grass.
Calamagrostis sylvatica breviseta. Gray.
   In the Black Hills.

Calamagrostis confinis, (Wild.) Nutt.   Bog Reed-grass.
   In the Minnesota and Sioux valleys.

Calamagrostis neglecta, (Ehrn.) Gaertn.   Narrow Reed-grass.
Calamagrostis stricta. Beauv.
   In the Sioux valley, near Brookings.

Calamagrostis Montanensis, Scrib.   Montana Reed-grass.
   Occasional in the Sioux and James valleys.

Calamovilfa longifolia, (Hook.) Hack.   Long-leaved Reed-grass.
Calamagrostis longifolia. Hook.
   In dry soils throughout the state.

Avena Striata, Michx.   Purple Oats.
   In the Black Hills.

Avena fatua, L.   Wild Oats.
   Introduced in the Minnesota and Sioux valleys.

Arrhenatherum Elatius, (L.) Beauv.   Oat-grass.
   Escaped from cultivation near Brookings.

Danthonia spicata, Beauv. Wild Oat-grass.
In the Black Hills.

Spartina Cynosuroides, (L.) Willd. Tall Marsh-grass.
In swamps and streams throughout the state.

Spartina gracilis, Trin. Inland Cord-grass.
In alkaline soils along Cheyenne river in the Bad Lands and along Lake Traverse.

Schedonnardus paniculatus, (Nutt.) Trelease. Schedonnardus.
Found occasionally along trails from the Missouri river to the Black Hills, also in the southern part of the state east of the river.

Bouteloua hirsuta, Lag. Hairy Mesquite-grass.
In dry soils throughout the state, but much less common than the next.

Bouteloua oligostachya, (Nutt.) Torr. Mesquite-grass.
Very abundant throughout the state; commonly called "False Buffalo Grass."

Bouteloua Curtipendula, (Michx.) Torr. Racemed Bouteloua.
Common throughout the state.

Beckmannia erucaeformis, (L.) Host. Beckmannia.
In wet places throughout the state.

Bulbilis dactyloides, (Nutt.) Raf. Buffalo-grass.
Throughout the state. It is rapidly disappearing in the eastern and southern part of the state.

Munroa squarrosa, (Nutt.) Torr. Munro's grass.
On the dry plain from the Missouri valley westward.

Phragmites Phragmites, (L.) Karst. Reed-grass.
In swamps and along the edges of streams from the Missouri river eastward.

Diplachne fascicularis, (Lam.) Beauv. Salt-meadow Diplachne.
In alkaline Marshes east of the Missouri river.

Eragrostis pilosa, (L.) Beauv.   Tufted Eragrostis.
Sparingly introduced into the eastern and southern part of the state.

Eragrostis purshii, Schrad.   Pursh's Eragrostis.
In the eastern part of the state.

Eragrostis Major, Host.   Strong-scented Eragrostis.
Common east of the Missouri, rare in the Black Hills.

Eragrostis hypnoides, (Lam.) B. S. P.   Creeping Eragrostis.
Common on sandy shores east of the Missouri river.

Eatonia obtusata, (Michx.) Gray.   Blunt-scaled Eatonia.
East of the Missouri and in the Black Hills.

Eatonia Pennsylvanica, (Dc.) Gray.   Pennsylvanian Eatonia.
East of the Missouri and in the Black Hills.

Koeleria cristata, (L.) Pers.   Koeleria.
Common from the Missouri river westward to the Black Hills.

Catabrosia aquatica, (L.) Beauv.   Waterwhirl-grass.
In swamps in the Black Hills.

Distichlis spicata, (L.) Greene.   Marsh Spike-grass.
Common in low alkaline soils throughout the state.

Dactylis glomerata, L.   Orchard-grass.
Sparingly introduced in pastures in the extreme eastern part of the state and in the Black Hills.

Poa compressa, L.   Wire-grass.
In dry places east of the Missouri river.

Poa pratensis, L.   Kentucky Blue-grass.
Introduced into meadows and lawns east of the river; probably native in the Black Hills.

Poa pseudopratensis, Scrib. & Ryd.   Prairie Meadow-grass.
Found near Hot Springs, in the Black Hills.

Poa trivialis, L. Rough Meadow-grass.
Sparingly introduced in the vicinity of Brookings.

Poa flava, L. False Red-top.
In swampy places in the extreme eastern part of the state; the Sioux and Little Minnesota valleys.

Poa nemoralis, L. Wood Meadow-grass.
On moist banks in the Sioux and James river valleys and in the Black Hills.

Poa debilis, Torr. Weak Spear-grass.
In wooded ravines in the Minnesota valley.

Poa alsodes, Gray. Grove Meadow-grass.
In damp woods in the Little Minnesota valley and in the Black Hills.

Poa arida, Vasey. Prairie Spear-grass.
In the Sioux and James valleys.

Poa alpina, L. Alpina Spear-grass.
Near Hot Springs, in the Black Hills.

Poa Buckleyana, Nash. Buckley's Spear-grass.
In dry soils near Hot Springs.

Poa laevis, Vasey. Smooth Poa.
Extends from the Missouri valley to the Black Hills.

Poa fendleriana, (Steud.) Vasey. Fendler's Poa.
In the Black Hills.

Poa nevedensis, Vasey. Nevada Poa.
In the Black Hills.

Poa annua, L. Annual Meadow-grass.
Elk Canon, in the Black Hills.

Panicularia nervata, (Willd.) Kuntze. Nerved Manna-grass.
Glyceria nervata. Trin.
In the Sioux and Little Minnesota valleys, and in the Black Hills.

Panicularia Americana, (Torr.) McM. Reed Meadow-grass.
Glyceria grandis, S. Wats.
Occasional in the Sioux valley and in the Black Hills.

Panicularia fluitans, (L.) Kuntz. Floating Meadow-grass.
Glyceria fluitans. R. B.
In shallow water in the Sioux valley.

Festuca octoflora, Walt. - Slender Fescue.
In dry sandy soil from the James valley to the Black Hills.

Festuca ovina, L. Sheep's Fescue.
Grows in bunches on dry prairies in the Black Hills.

Festuca Nutans, Willd. Nodding Fescue.
On shady bluffs in the Sioux and Little Minnesota valleys.

Bromus ciliatus, L. Fringed Brome.
In thickets in the Sioux, James and Little Minnesota valleys, and in the Black Hills.

Bromus Kalmii, Gray. Kalm's Chess.
In the Black Hills.

Bromus pimpellianus. Scribner.
In the Black Hills.

Agropyron repens, (L.) Beauv. Cough-grass, "Quack"-grass.
A most troublesome weed in cultivated fields east of the Missouri river.

Agropyron violaceum, (Horn.) Vasey. Purplish wheat-grass.
Found occasionally on high ground, from the Missouri valley eastward. Specimens collected by Dr. Rydberg in the Black Hills were doubtfully referred to *Agropyron violaceum majus.*

Agropyron spicatum, (Pursh.) Scrib. & Smith. Western Wheat-grass. Alkali-grass.

Throughout the state. In the James and Missouri valleys it forms a great portion of the forage. It prefers a damp, heavy, somewhat alkaline soil. It is not common on the prairies in the eastern part of the state, but is becoming more abundant where the land is broken.

Agropyron tenerum, Vasey. Slender Wheat-grass.

Common on dry prairies throughout the state.

Agropyron strygosum, Beauv. Rough Wheat-grass.

In sterile soil along Indian creek in the Bad Lands.

Agropyron caninum, (L.) R. & S. Awned Wheat-grass.

Common in the Sioux valley and in the Black Hills, probably throughout the state.

Hordeum Jubatum, L. Squirrel-tail-grass.

A very common and troublesome weed in all waste places throughout the state.

Elymus striatus, Willd. Slender wild rye.

On banks of streams from the Missouri valley eastward, and in the Black Hills.

Elymus virginicus, L. Terrell-grass; wild rye.

Common along streams throughout the state.

Elymus canadensis, L. Nodding wild rye.

Common along streams throughout the state. Most of the material from the arid regions west of the Missouri river and in the Black Hills belongs to the variety Glaucifolius Torr.

Elymus Macounii, Vasey. Macoun's wild rye.

In the Little Minnesota valley.

Elymus Elymoides, (Raf.) Swezey. Long-bristled wild rye.

From the Missouri river westward; rare.

Elymus dasystachys, Trin. Western wild rye.

In the Black Hills.

CYPERACEÆ.—The Sedge Family.

Cyperus inflexus, Muhl. Awned Cyperus.
Cyperus Aristatus, of Manuals.
In damp sandy soil from the Missouri river eastward.

Cyperus Schweinitzii, Torr. Schweinitz Cyperus.
In low, moist ground from the Missouri valley eastward.

Cyperus acuminatus, Torr. & Hook. Short-pointed Cyperus.
In wet meadows throughout the state.

Cyperus erythrorhizos, Muhl. Red-rooted Cyperus.
In the Sioux and James valleys.

Cyperus strigosus, L. Straw-colored Cyperus.
In wet meadows in the little Minnesota valley.

Cyperus ovularis, (Michx.) Torr. Globose Cyperus.
In the Sioux and Little Minnesota valleys.

Eleocharis Englemanni. Steud. Englemann's Spike-rush.
In wet soil from the Missouri valley eastward. There are several specimens in the College herbarium labelled *E. Ovata*. The plants all have the pointed spike, the low broad tubercle covering the top of the achene and the short bristles of E. Englemanni.

Eleocharis palustris, (L.) R. & S. Creeping Spike-rush.
In swamps and ponds in the Black Hills, James, Sioux and Little Minnesota valleys; and in the southern part of the Missouri valley.

Eleocharis acicularis, (L.) R. & S. Needle Spike-rush.
In wet soils from the Missouri valley eastward and probably throughout the state.

Eleocharis acuminata, (Muhl.) Nees. Flat-stemmed Spike-rush.
In the Black Hills near Hot Springs.

Eleocharis intermedia, (Muhl.) Schuttes. Matted Spike-rush.
In springy marshes in the Sioux valley, near Brookings; rare.

Scirpus pauciflorus, Lightf. Few-flowered Club-rush.
In the Black Hills, near Custer.

Scirpus debilis, Pursh. Weak-stalked Club-rush.
In the Missouri valley, in Potter and Walworth counties.

Scirpus Americanus, Pers. Three-squares.
Scirpus pungens, Vahl.
In brackish, or fresh water swamps throughout the state; rare in the Black Hills.

Scirpus lacustris, L. Great Bulrush.
In lakes and ponds throughout the state.

Scirpus fluviatilis, (Torr.) Gray. River Bulrush.
On sandy shores of lakes and ponds, and along slow streams from the Missouri valley eastward throughout the state. In one locality "between 600 and 1,000 acres came up in June, 1894, and yielded from 16 to 25 bushels per acre of seed, which was used for feed for stock and chickens." .

Scirpus atrovirens, Muhl. Dark-green Bulrush.
In swamps in the Sioux and Little Minnesota valleys.

Scirpus atrovirens pallidus, Britton. Pale Sedge.
In the Black Hills.

Scirpus cyperinus, (L.) Kunth. Wood-grass.
In the Black Hills, near Custer.

Eriophorum polystachyon, L. Tall Cotton-grass.
In springy bogs in the Sioux valley. Watertown, Toronto, Elkton.

Eriophorum gracile, Kock. Slender Cotton-grass.
Collected in a springy bog in the extreme eastern part of the state, near Elkton.

Carex lupuliformis, Sartwell. Hop-like Sedge.
In swamps and lake margins in the Little Minnesota and Sioux valleys; forming a considerable of the forage on low ground.

Carex festiva, Dewey. Festival Sedge.
Very rare, in the Black Hills.

Carex utriculata, Boott. Bottle Sedge.
In marshes in the Little Minnesota and Sioux valleys and in the Black Hills.

Carex hystricina, Muhl. Porcupine Sedge.
In springy swamps in the extreme eastern part of the Sioux valley.

Carex monile, Tuckerm. Necklace Sedge.
In a springy bog near Elkton; the extreme eastern part of the Sioux valley.

Carex retrorsa, Schwein. Retrorse Sedge.
In wet meadows in the Sioux and James valleys.

Carex pseudo-cyperus, L. Cyperous-like Sedge.
In low swails and margins of ponds in the Sioux valley.

Carex trichocarpa, Muhl. Hairy-fruited Sedge.
In lakes and marshes in the Little Minnesota, the Sioux, and the southern part of the Missouri valley.

Carex aristata, R. Br. Awned Sedge.
In swamps in the Sioux valley.

Carex Houghtonii, Torr. Houghton's Sedge.
On dry banks in the Sioux valley near Brookings.

Carex lanuginosa, Michx. Woolly Sedge.
In springy swamps in the Little Minnesota, the Sioux and the southern part of the James and Missouri valleys.

Carex filiformis, L. Slender Sedge.
In the Sioux valley near Brookings.

Carex stricta, Dewey. Tussock Sedge.
 In low meadows in the Sioux and James valleys, and the southern part of the Missouri valley.

Carex Haydenii, Dewey. Hayden's Sedge.
 In sloughs in the Little Minnesota, Sioux and the James valleys.

Carex Nebraskensis, Dewey. Nebraska Sedge.
 In the Black Hills near Custer.

Carex longirostris, Torr. Long-beaked Sedge.
 On damp shady banks in the Little Minnesota and Sioux valleys and in the Black Hills.

Carex Assiniboinensis, W. Boott. Assiniboia Sedge.
 On damp shady banks in the Little Minnesota, and the northern part of the Sioux valley.

Carex capillaria, L. Hair-like Sedge.
 On moist shady banks bordering streams in the Little Minnesota valley.

Carex grisea, Wahl. Gray Sedge.
 In low woods in the Sioux valley near Brookings.

Carex tetanica, Schk. Wood Sedge.
 In low ground in the Sioux valley.

Carex Meadii, Dewey. Mead's Sedge.
 In wet meadows in the Sioux valley near Brookings.

Carex laxiflora blanda, (Dewey.) Boott. Loose-flowered Sedge.
 In damp woods near lakes in the Little Minnesota and Sioux valleys.

Carex aurea, Nutt. Golden-fruited Sedge.
 In the Black Hills, near Lead.

Carex Richardsonii, R. B. Richardson's Sedge.
 In the Sioux and James valleys and in the Black Hills, not abundant.

Carex pedicellata, (Dewey.) Britton. Fibrous-rooted Sedge.
On bluffs in the Little Minnesota valley; rare.

Carex Pennsylvanica, Lam. Pennsylvania Sedge.
Very abundant in dry soils, both open and shaded. From the Missouri valley eastward, and in the Black Hills, probably throughout the state.

Carex varia, Muhl. Emmons' Sedge.
In dry soils in the Black Hills.

Carex filifolia, Nutt. Thread-leaved Sedge.
In dry soil throughout the state, more abundant in the central and western part.

Carex stenophylla, Wahl. Involute-leaved Sedge.
In dry soil throughout the state. This plant and C. Pennsylvanica form not an inconsiderable amount of early forage.

Carex Marcida, Boott. Clustered Field Sedge.
In the Little Minnesota valley and in the Black Hills.

Carex gravida, Bailey. Heavy Sedge.
On low ground in the Sioux valley.

Carex vulpinoidea, Michx. Fox Sedge.
In the Sioux, James and Missouri valleys.

Carex Sartwellii, Dewey. Sartwell's Sedge.
In the Sioux and the Little Minnesota valleys.

Carex tenella, Schk. Soft-leaved Sedge.
Near Sylvan Lake in the Black Hills.

Carex rosea, Schk. Stellate Sedge.
On damp wooded bluffs of Bigstone Lake in the Little Minnesota valley.

Carex sterilis, Willd. Little Prickly Sedge.
In the Sioux valley near White.

Carex siccata, Dewey. Hillside Sedge.
In the Sioux, James and Missouri valleys and in the Black Hills.

Carex tribuloides, Wahl. Blunt Broom Sedge.
In the Sioux valley.

Carex tribuloides Bebbii, Bailey.
Occurs in the Black Hills.

Carex foenea, Willd. Hay Sedge.
In the Sioux valley near Brookings; rare.

Carex Deweyana, Schwein. Dewey's Sedge.
Occurs rather rarely in the Black Hills and in the Sioux valley.

Carex straminea, Willd. Straw Sedge.
In dry soils in the Sioux and Little Minnesota valleys.

Carex festucaceae, Willd. Fescue Sedge.
In the Sioux valley near Brookings.

Carex Bicknellii, Britton. Bicknell's Sedge.
Carex Straminea Crawei, Boott.
Near Hot Springs in the Black Hills; rare.

Carex sychnocephala, Carey. Dense Long-beaked Sedge.
In low meadows in the Sioux valley, near Brookings, and in the Little Minnesota valley near outlet of Bigstone Lake.

ARACEÆ.—The Arum Family.

Arisæma triphyllum, (L.) Torr. Indian Turnip.
On moist shady banks bordering lakes and streams in the Little Minnesota and the Sioux valleys.

Acorus calamus, L. Sweet Flag.
Collected in two localities in the extreme eastern part of the state.

LEMNACEÆ.—Duckweed Family.

Spirodella polyrhiza, (L.) Schleid. Greater Duckweed.
In ditches, ponds and lakes throughout the state.

Lemna trisulca, L. Star Duckweed.
In ponds and pools; so far it has been collected only from the Missouri river eastward.

Lemna minor, L. Lesser Duckweed.
In ponds, lakes and stagnant water throughout the state.

COMMELINACEÆ.—Spiderwort Family.

Tradescantia virginiana, L. Spiderwort.
Abundant in low moist ground throughout the state.

PONTEDERIACEÆ.—Pickerelweed Family.

Heteranthera limosa, (Sw.) Willd. Smaller Mud Plantain.
In muddy ponds near Dell Rapids, in the Sioux valley and at various points in the Missouri valley.

Heteranthera dubia, (Jacq.) McM. Water Star-grass.
Schollera graminea, Gray.
In clear water from the Missouri eastward.

JUNCACEÆ.—Rush Family.

Juncus effusus, L. Bog Rush.
Common in low ground in the Sioux valley.

Juncus bufonius, L. Toad Rush.
In the Black Hills; rare.

Juncus tenuis, Willd. Slender Rush.
Common in the Sioux and the James river valleys and in the Black Hills.

Juncus Vaseyi, Englm. Vasey's Rush.
Near Hot Springs, in the Black Hills.

Juncus longistylis, Torr. Long-styled Rush.
Near Lead City, in the Black Hills.

Juncus nodosus, L. Knotted Rush.
In the Sioux, the Little Minnesota, and the James valleys, and the Black Hills.

Juncus Torreyi, Coville. Torrey's Rush.
In the Sioux, Minnesota, James and Missouri valleys, and in the Black Hills.

Juncus Xiphioides Montanus, Englm. The Mountain Sedge.
A rare plant near Custer, in the Black Hills.

Juncoides comosum, (Meyer.) Sheldon. The Hairy Rush.
In the Black Hills; rare.

MELANTHACEÆ.—Bunch-Flower Family.

Zygadenus elegans, Pursh. Glaucous Zygadenus.
In low prairies probably throughout the state, the Little Minnesota, Sioux, James and Missouri valleys, and in the Black Hills.

Zygadenus venosus, S. Wats. Poisonous Zygadenus.
In the Black Hills, near Hot Springs, Rapid City, etc.

Uvularia grandiflora, J. E. Smith. Large-flowered Bellwort.
In damp shady ravines in the Little Minnesota valley.

LILIACEÆ.—The Lily Family.

Leucocrinum montanum, Nutt. Leucocrinum.
Common in the Black Hills, Custer, Rapid City and the adjacent plains.

Allium tricoccum, Ait. Wild Leek.
In damp deep ravines in the Minnesota valley.

Allium cernuum, Both. Nodding Wild Onion.
In the Sioux valley and the Black Hills, abundant.

Allium stellatum, Kerr. Prairie Wild Onion.
Common in dry soils in the Sioux valley and in the Black Hills.

Allium Canadensis, L. Meadow Garlic.
Common in low prairies in the Sioux valley.

Allium Nuttallii, S. Wats. Nuttall's Wild Onion.
In dry soils in the southern part of the James and Missouri valleys.

Allium reticulatum, Don. Fraser's Wild Onion.
On dry prairies from the Missouri river eastward and in the Black Hills.

Allium Geyeri, Wats. Geyer's Wild Onion.
In the Black Hills.

Lilium umbellatum, Pursh. Western Red Lily.
In low prairies in the Little Minnesota valley and in the Black Hills.

Fritillaria atropurpurea, Nutt. Purple Fritillaria.
In the Bad Lands region in the south central part of the state.

Calochortus Nuttallii, T. & G. Nuttall's Mariposa Lily.
In the Black Hills.

Calochortus Gunnisonii, S. Wats. Gunnison's Mariposa Lily.
In the Black Hills and in the Bad Lands.

Yucca glauca, Nutt. Bear-grass, Indian Soapweed.
Common in the dry soils, especially on bluffs from the Missouri valley to the Black Hills.

CONVALLARIACEÆ.—Lily of the Valley Family.

Asparagus, Officinalis, L. Asparagus.
Escaped in fields and timber claims in a few places; Brookings, Yankton and Dell Rapids.

Vagnera racemosa, (L.) Morong. Wild Spiknard.
Smilacina racemosa.
On damp, shady banks in the Sioux and Little Minnesota valleys.

Vagnera stellata, (L.) Morong.   Star-flowered Solomon's Seal.
Smilacina Stellata, Nutt.
In the Little Minnesota, Sioux, James and Missouri valleys, and in the Black Hills.

Vagnera amqlexicaulis, (Nutt.) Greene.   Western Solomon's Seal.
Smilacina amplexicaulis, Nutt.
In the Black Hills.

Unifolium Canadense, (Desf.) Greene.   False Lily-of-the-Valley.
Majanthemum Canadense, Desf.
In shady woods in the Minnesota valley and in the Black Hills.

Diosporum trachycarpum, (S. Wats.) B. & H.   Rough-fruited Diosporum.
Prosartes trachycarpum, S. Wats.
In shady places in the Black Hills.

Streptopus amplexifolius, (L.) Dec.   Clasping-leaved Twist-foot.
Near Sylvan Lake, in the Black Hills.

Polygonatum Commutatum, (R. & S.) Dietr.   True Solomon's Seal.
In moist Woods from the Missouri valley eastward and in the Black Hills.

Trillium erectum, L.   Ill-scented Wake Robin.
In deep ravines in the Minnesota valley. Fruiting specimens were also collected in the same locality with the last which agree with *T. Grandiflorum,* but as no flowers were collected this is not reported as a certainty.

SMILACACEÆ.—The Smilax Family.

Smilax herbacea, L.   Carrion Flower.
In woods and thickets throughout the state.

Smilax hispida, Muhl. Hispid Greenbrier.
 In the southern part of the Sioux valley from Sioux Falls southward and in the Southern Missouri valley to Running Water.

AMARYLLIDACEÆ.—Amaryllis Family.

Hypoxis hirsuta, (L.) Coville. Star-grass.
 Common on prairies in the Minnesota and Sioux valleys.

IRIDACEÆ.—The Iris Family.

Iris Missouriensis, Nutt. Western Flag.
 In wet soils throughout the Black Hills.

Sisyrinchium angustifolium, Mill. Pointed Blue-eyed Grass.
 Common from the Missouri valley eastward and in the Black Hills.

ORCHIDACEÆ.—The Orchid Family.

Cyprepedium candidum, (Willd.) Small White Ladies' Slipper.
 In low, damp meadows in the Minnesota and Sioux valleys.

Cyprepedium hirsutum, Mill. Large Yellow Ladies' Slipper.
 In the deep, shaded ravines of the Minnesota valley.

Cyprepedium parviflorum, Salisb. Small Yellow Ladies' Slipper.
 In woods in the Minnesota valley and in the Black Hills.

Habenaria hyperborea, (L.) Tall Green Orchis.
 In bogs in the Little Missouri valley and in the Black Hills.

Habenaria bracteata, (Wild.) R. Br. Long-bracted Orchis.
  In damp woods in the Little Minnesota valley and in the Black Hills.

Habenaria leucophæa, (Nutt.) A. Gray. Prairie White-fringed Orchis.
  In low prairies in the Sioux valley near Brookings; rare.

Gyrostachys Romanzoffiana, (Cham.) MacM. Hooded Ladies' Tresses.
  Spiranthes Romanzoffiana, (Cham.)
  In low ground in the Sioux valley and in the Black Hills.

Peramium repens, (L.) Salisb. Lesser Rattlesnake Plaintain.
  Goodyera repens, R. Br.
  In the Black Hills.

Corallorhiza Corollorhiza, (L.) Karst. Early Coral-root.
  In woods in the Little Minnesota valley and in the Black Hills.

Corallorhiza multiflora, Nutt. Large Coral-root.
  On shady banks in the Black Hills.

DICOTYLEDONES.

JUGLANDACEÆ.—Walnut Family.

Juglans nigra, L. Black Walnut.
  Occurs native only in the southeastern part of the state; Union county.

SALICACEÆ.—Willow Family.

Populus balsamifera, L. Balsam Poplar.
In the deep, wooded ravines of the Minnesota valley; rare. It is reported also from near Sioux Falls, but no specimens have been seen.*

Populus angustifolia, James. Narrow-leaved Cottonwood.
In the Black Hills.

Populus acuminata, Rydberg. Black Cottonwood.
Near Hot Springs in the Black Hills.

Populus tremuloides, Michx. American Aspen.
On the dry bluffs and in the spring swamps in the Minnesota valley, in the higher altitudes in the Black Hills, and in isolated patches between the Missouri river and the Black Hills. It is reported also from the Sioux valley, but no specimens have been seen.

Populus deltoides, Marsh. Cottonwood.
Around lakes and bordering streams throughout the state; common.

Salix nigra, Marsh. Black Willow.
A shrub or small tree, common along streams in the Minnesota, Sioux and James valleys.

Salix fluviatilis, Nutt. Sand-bar Willow.
Salix longifolia, Gray.
A small, slender shrub, along streams and lakes throughout the state; the commonest of the willows.

Salix Bebbiana, Sarg. Bebb's Willow.
Salix rostrata, Richards.
In the Minnesota valley and in the Black Hills.

Salix humilis, Marsh. Prairie Willow.
Common in the Minnesota region; it is usually found on the wooded bluffs or the edge of the open prairie, acting as the advance guard of the wooded formations.

---

*Williams, Bulletin 43, U. S. Exp. Sta., S. D.; 105; 1895.

Salix discolor, Muhl.   Glaucous Willow.
  In cold swamps in the Minnesota valley and in the Black Hills.

Salix cordata, Muhl.   Heart-leaved Willow.
  Along streams throughout the state.

Salix balsamifera, (Hook.) Barrett.   Balsam Willow.
  In the cold spring swamps in the Minnesota region.

Salix adenophylla, (Hook.)   Furry Willow.
  A single specimen was collected on the shores of one of the numerous lakes found in the coteaus in the northeastern part of the state, which is doubtfully referred to this species.

Salix myrtilloides, L.   Bog Willow.
  In the boggy swamps at the head of the coulies in the Minnesota region.

BETULACEÆ.—The Birch Family.

Ostrya virginica, (Mill.)   Iron-wood.
  On wooded bluffs in the Minnesota region, and among the foot-hills in the Black Hills.

Corylus Americana, Walt.   Hazelnut.
  In thickets and open woods in the Minnesota and Sioux regions, and in the Black Hills.

Corylus rostrata, Ait.   Beaked Hazelnut.
  In the Black Hills, not as common as the last.

Betula papyrifera, Marsh.   Canoe Birch.
  Common in the Black Hills.

Betula occidentalis, Hook.   Western Red Birch.
  In the Black Hills; common.

Betula glandulosa, Michx.   Scrub Birch.
  A low shrub, abundant in many places in the Black Hills.

FAGACEÆ.—Beech Family.

Quercus macrocarpa, Michx.   Burr Oak.

In rich, open woods and on dry bluffs, in the vicinity of streams or lakes throughout the state. The White Oak (*Quercus Alba*) has been reported for this state, but wherever specimens were received or the locality visited, the tree proved to be one of the numerous forms of the Burr Oak. It seems very doubtful if the white oak occurs in this state.

ULMACEÆ.—The Elm Family.

Ulmus Americana, L.   White Elm.

Along streams and lakes throughout the state.

Ulmus fulva, Michx.   Slippery Elm.

Extends up the Sioux river to Sioux Falls, and up the Missouri river nearly to Chamberlain. A few trees were also found around Buffalo Lake in the northeastern part of the state, just west of the head waters of the Little Minnesota.

Celtis occidentalis, L.   Hackberry.

In the vicinity of lakes and streams throughout the state.

MORACEÆ.—The Mulberry Family.

Morus rubra, L.   Red Mulberry.

This tree is found naturally only in the extreme southeastern county of the state, along the Sioux river; Elk Point.

Humulus Lupulus, L.   Hops.

In thickets bordering streams and lakes throughout the state.

Cannabis sativa, L.   Hemp.

Sparingly introduced into the state from the Missouri river eastward.

URTICACEÆ.—Nettle Family.

Urtica gracilis, Ait. Slender Nettle.
In thickets and low ground throughout the state.

Urticastrum divaricatum, (L.) Kuntze. Wood Nettle.
Laportea Canadensis, Gaud.
In low, rich woods from the Missouri river eastward.

Adicea Pumila, (L.) Raf. Clearweed.
Pilea pumila, Gray.
In damp, shady ground in the Minnesota and Sioux valleys.

Parietaria Pennsylvanica, Muhl. Pennsylvanian pellitory.
In shady woods from the Missouri eastward, and in the Black Hills.

SANTALACEÆ.—Sandal-wood Family.

Comandra umbellata, (L.) Nutt. Bastard Toad-flax.
On dry, gravelly bluffs from the Missouri valley eastward.

Comandra pallida, A. Dc. Pale Comandra.
From the Missouri river westward throughout the state, including the Black Hills.

POLYGONACEÆ.—Buckwheat Family.

Eriogonum annum, Nutt. Annual Eriogonum.
In the Black Hills, and also extending over most of the plains region from the Missouri valley westward.

Eriogonum multiceps, Nees. Branched Eriogonum.
On the dry plains from the Missouri river westward, and in the Black Hills.

Eriogonum pauciflorum, Pursh. Few-flowered Eriogonum.
In the Black Hills, and in the surrounding plain region.

Eriogonum flavum, Nutt.   Yellow Eriogonum.
In the Black Hills.

Rumex acetosella, L.   Sheep Sorrel.
A European weed introduced into pastures and timber claims, etc., in the Sioux valley and in the Black Hills.

Rumex venosus, Pursh.   Veined Dock.
From the bluffs of the Missouri river westward, including the Black Hills.

Rumex altissimus, Wood.   Peach-leaved Dock.
Common along streams and in wet ground from the Missouri valley eastward.

Rumex salicifolius, Weinm.   Pale Dock.
In low ground near lakes and streams throughout the state.

Rumex Britannica, L.   Great-water Dock.
In the Minnesota and Sioux valleys.

Rumex occidentalis, S. Wats.   Western Dock.
On the plains west of the Missouri river, and in the Black Hills.

Rumex crispus, L.   Curled Dock.
Sparingly introduced into the state from the Missouri river eastward, and in the Black Hills.

Rumex persicarioides, L.   Golden Dock.
Abundant on damp, shady shores from the Missouri valley eastward.

Polygonum viviparum, L.   Alpine Bistort.
In damp, mossy places in the Black Hills.

Polygonum amphibium, L.   Water persicaria.
In shallow water in the Minnesota, the Sioux and the James valleys.

Polygonum Hartwrightii, Gray.   Hartwright's Persicaria.
In the Sioux valley; rare.

Polygonum emersum, (Mich.) Britton.   Swamp Persicaria.
: In swamps and edges of ponds throughout the state.

Polygonum lapathifolium, L.   Dock-leaved Persicaria.
: In waste places throughout the state; rare west of the Missouri, except in the Black Hills.

Polygonum lapathifolium incanum, (Schmidt.) Kock.
: Same range as the type.

Polygonum Pennsylvanicum, L.   Pennsylvania Persicaria.
: In moist soil from the Missouri valley eastward.

Polygonum Persicaria, L.   Ladies' Thumb.
: Common in waste places from the Missouri valley eastward, and in the Black Hills.

Polygonum hydropiper, L.   Smart-weed.
: In moist places in the Minnesota and Sioux valleys.

Polygonum punctatum, Ell.   Water Smart-weed.
: *Polygonum acre*, H. B. K.
: In cold swamps in the Minnesota valley.

Polygonum aviculare, L.   Knot-grass.
: In waste ground throughout the state.

Polygonum littorale, Link.   Shore Knot-weed.
: In waste places in the Black Hills.

Polygonum erectum, L.   Erect Knot-weed.
: In dry soils throughout the state.

Polygonum ramosissimum, Michx.   Bushy Knot-weed.
: A common weed in dry soils, throughout the state.

Polygonum camporum, Meisn.   Prairie Knot-weed.
: On prairies from the Missouri river eastward.

Polygonum tenue, Michx.   Slender Knot-weed.
: In ground that has been broken from the Missouri river eastward.

Polygonum Douglassii, Greene. Douglas Knot-weed.
   In the Black Hills and near Sand Lake in the James valley.

Polygonum Convolvulus, L. Black Bind-weed.
   A most troublesome weed in cultivated ground throughout the state.

Polygonum scandens, L. Climbing False Buckwheat.
   In thickets from the Missouri river eastward.

Polygonum Sawachense, Small. Western Persicaria.
   Near Custer, in the Black Hills.

CHENOPODIACEÆ. — The Goosefoot Family.

Chenopodium album, L. Lamb's-quarter. Pigweed.
   A weed naturalized in waste places throughout the state.

Chenopodium glaucum, L. Oak-leaved Goosefoot.
   A weed naturalized; it prefers low, alkaline soils; in the Minnesota and Sioux valleys.

Chenopodium leptophyllum, (Moq.) Nutt. Narrow-leaved Goosefoot.
   In the Black Hills.

Chenopodium leptophyllum oblongifolium, Wats. The Oblong-leaved Chenopod.
   Near the Cheyenne river, east of the Black Hills.

Chenopodium Boscianum, Moq. Bosc's Goosefoot.
   In woods in the Minnesota region.

Chenopodium Fremontii, S. Wats. Fremont's Goosefoot.
   In damp woods in the Black Hills.

Chenopodium Fremontii incanum, S. Wats.
   In the Black Hills.

Chenopodium hybridum, L. Maple-leaved Goosefoot.
   In open woods in the Minnesota and Sioux valleys, and in the Black Hills.

Chenopodium rubrum, L. Red Goosefoot.
In low, alkaline soils in the Sioux, Minnesota and James valleys.

Blitum capitatum, L. Strawberry Blight.
Chenopodium capitatum, (L.) Wats.
In the Black Hills.

Cycloma atriplicifolium, (Spreng.) Coult. Cycloma.
Cycloma platyphyllum, Moq.
In the foot-hills of the Black Hills, and the adjacent plains region; Rapid City, etc.

Monolepos Nuttaliana, (R. & S.) Greene. Monolepis.
Monolepis chenopodioides, Moq.
In alkaline soils in the Black Hills, and from there eastward to the Missouri river.

Atriplex hastata, L. Halbert-leaved Orache.
In low, alkaline places throughout the state.

Atriplex argentea, Nutt. Silver Orache.
In alkaline soils from the Missouri river westward.

Atriplex canescens, (Pursh.) James. Bushy Atriplex.
Abundant in alkaline soils from the Missouri river westward; not yet reported from the Black Hills.

Eurotia lanata, (Pursh.) Moq. White Sage.
In the Bad Lands, just east of the Black Hills.

Corispermum hysopifolium, L. Bug-seed.
In the Bad Lands, east of the Black Hills.

Salicornia herbacea, L. Slender Glasswort.
In low, alkaline meadows in the Minnesota region near Wilmot, and in the James valley near Iroquois.

Dondia depressa, (Pursh.) Britton. Western Blight.
In alkaline soils with the last throughout the state.

Salsola tragus, L. Russian Thistle.
In waste places throughout the state, but most abundant in the James and the Missouri valleys. A very

troublesome weed in waste places, but easily destroyed by cultivation. When young and tender it is readily eaten by sheep.

AMARANTHACEÆ.—Amaranth Family.

Amaranthus retroflexus, L. Rough Pigweed.
An introduced weed in waste soil throughout the state, but not common in the central and western part.

Amaranthus hybridus, L. Slender Pigweed.
A naturalized weed in waste places from the Missouri river eastward, less common than the last.

Amaranthus blitoides, S. Wats. Prostrate Amaranth.
In cultivated and waste fields from the Missouri river eastward.

Amaranthus græcizans, L. Tumbleweed.
A common weed in cultivated ground throughout the state.

Acnida tamariscina, (Nutt.) Wood. Western Water-hemp.
Occasional in swamps and low places, which are somewhat brackish, from the Missouri river eastward.

Acnida tamarascina tuberculata, (Moq.) Uline & Bray. Tubercaled Water-hemp.
Same range as the last.

NYCTAGINACEÆ.—Four-o'clock Family.

Allionia nyctaginea, Michx. Heart-leaved Umbrella-wort.
In thickets throughout the state.

Allionia albida, Walt. Pale Umbrellawort.
In the Black Hills.

Allionia hirsuta, Pursh. Hairy Umbrellawort.
In dry soil throughout the state.

Allionia linearis, Pursh. Narrow-leaved Umbrellawort.
    In dry soil from the Missouri valley westward.
Abronia fragrans, Nutt. White Abronia.
    In the Bad Lands, east of the Black Hills.

PORTULACACEÆ.—Purslane Family.

Talinum teretifolium, Pursh. Fame-flower.
    On dry, rocky hills in the Sioux valley near Dell Rapids, and in the Black Hills.
Claytonia perfoliata amplectens, Greene. Spanish Lettuce.
    In the Black Hills.
Portulaca oleracea, L. Pursley. Purslane.
    An introduced weed in cultivated grounds throughout the state.

CARYOPHYLLACEÆ.—The Pink Family.

Silena Noctiflora, L. Night-flowering Catchfly.
    An introduced weed, occurring occasionally in the Sioux valley.
Lychinis alba, Mill. White Champion.
    Sparingly introduced near Brookings.
Lychinis Drummondii, (Hook.) S. Wats. Drummond's Pink.
    In the Black Hills, Custer and Rapid City.
Saponaria officinalis, L. Soapwort, Bouncing Bet.
    Escaped from cultivation in the Sioux valley.
Vaccaria vaccaria, (L.) Britton. Cow-herd.
    Saponaria vaccaria, L.
    Occurs occasionally in the Sioux and Minnesota valleys, and in the Black Hills.
Alsine media, L. Chickweed.
    Stellaria media, Cyr.
    Sparingly introduced in the Sioux valley.

Alsine longifolia, (Muhl.) Britton. Long-leaved Stitchwort.
Stellaria longifolia, Muhl.
In the Minnesota and Sioux valley, and in the Black Hills.

Alsine borealis, (Bigel.) Britton. Northern Stitchwort.
Stellaria borealis, Bigel.
In cold bogs in the Minnesota and the extreme eastern part of the Sioux valley, Elkton; rare.

Agrostemma Githago, L. Corn Cockle.
In grain fields from the Missouri river eastward.

Silena acaulis, L. Moss Champion.
In the Black Hills, Rapid City.

Silena stellata, (L.) Ait. Starry Champion.
In woods in the southern part of the Sioux valley; Flandreau, Sioux Falls.

Silena vulgaris, (Moench.) Garcke. Bladder Champion.
An introduced weed in the Sioux valley near Brookings.

Silena antirrhina, L. Sleepy Catchfly.
Occasional in low prairies in the Minnesota, Sioux and James valleys, and in the Black Hills.

Cerastium longipedunculatum, Muhl. Powderhorn.
Abundant in moist shade in the Black Hills.

Cerastium brachypodium, (Englm.) Robinson. Short-stalked Chickweed.
From the Missouri river eastward, and in the Black Hills.

Cerastium arvense, L. Field Chickweed.
Same range as the last, but more abundant.

Cerastium arvense oblongifolium, (Torr.) Holl. & Brett.
Occasional in the Sioux, the James valley, and in the Black Hills.

Arenaria Hookerii, Nutt. Hooker's Sandwort.
In dry, rocky soil in the Black Hills, and the adjoining hills and buttes.

Arenaria verna, L. Vernal Sandwort.
In shady, rocky soil in the Black Hills.

Arenaria stricta, Michx. Rock Sandwort.
In sandy soil in the Black Hills.

Mœhringia lateriflora, (L.) Fenyl. Blunt-leaved Sandwort.
Arenaria lateriflora, L.
In rich, shady soil in the Minnesota valley, and in the Black Hills.

Spergula arvensis, L. Spurey.
A naturalized weed, sparingly introduced in the Sioux valley.

Paronychia Jamesii, T. & G. James' Whitlow-wort.
On dry soil in the Black Hills, and the surrounding plains.

NYMPHÆACEÆ.—Water Lily Family.

Nymphæa advena, Soland. Large Yellow Pond Lily.
Nuphar advena, R. Br.
In ponds and streams throughout the state, except in the dryer part of the plains region.

Nymphæa oderata, (Dryand.) Woods & Wood. Pond Lily.
Quite authentic reports have been received of the occurrence of this species in a tributary of the Sioux river southeast of Brookings, in the extreme eastern part of the state, but no specimens have been seen.

CERATOPHYLLACEÆ.

Ceratophyllum demersum, L. Hornwort.
One of the most abundant plants in ponds, lakes and slow streams, from the Missouri river eastward.

Ranunculaceæ.—The Crowfoot, or Buttercup Family.

Caltha palustris, L. Marsh Marigold.
In cold, springy swamps and low meadows in the Minnesota valley.

Actæa rubra, (Ait.) Willd. Red Baneberry.
In rich woods near lakes and streams in the Sioux valley, and in the Black Hills.

Actæa rubra arguta, (Nutt.) Greene. Western Baneberry.
Occurs in the Black Hills.

Actæa alba, (L.) Mill. White Baneberry.
In the Sioux valley with the last species.

Aquilegia Canadensis, L. Wild Columbine.
On damp, shady banks in the vicinity of lakes and streams in the Minnesota, Sioux, James, and the southern part of the Missouri valleys, and in the Black Hills.

Aquilegia Canadensis formosa, (Fisch.) Cooper.
Occurs rarely in the Black Hills.

Aquilegia brevistyla, Hook. Small-flowered Columbine.
On shady banks in the Black Hills.

Delphinium Carolinianum, Walt. Carolina Larkspur.
Common in open ground from the Missouri eastward.

Delphinium bicolor, Nutt. Mewzie's Larkspur.
A variable species, common in the Black Hills.

Delphinium urceolatum, Jacq. Tall Larkspur.
A single fragmentary specimen was received from Rapid City in the spring of 1898, which is very doubtfully referred to this species.

Aconitum Fischeri, Reich. Fisher's Monkshood. Wolfsbane.
In damp ravines in the Black Hills.

Anemone Caroliniana, Walt. Caroline Anemone.
On prairies, especially in low places from the Missouri eastward.

Anemone multifida, Poir. Red Wind Flower.
Occurs only in the Black Hills.

Anemone cylindrica, A. Gray. Long-fruited Anemone.
On prairies throughout the state.

Anemone virginiana, L. Tall Anemone.
In open woods in the Minnesota and Sioux valleys.

Anemone Canadensis, L. Canadian Anemone.
In low ground, especially in the vicinity of woods or thickets, from the Missouri eastward.

Pulsatilla hirsutissima, (Pursh.) Britton. Pasque Flower.
Anemone patens Nutalliana, Dc.
Throughout the state; especially abundant on sandy bluffs and hills from the Missouri river eastward; one of the earliest flowers that blooms on the open prairie. Quite severe losses occasionally occur to the sheep industry by the formation of "hair balls" in the stomach of sheep which have eaten too greedily of this plant. The trouble most often occurs early in the spring, before the grasses have made much growth.

Clematis Virginiana, L. Virginian Virgin's Bower.
Along streams and lakes in the Sioux, James and Minnesota valleys.

Clematis ligusticifolia, Nutt. Western Virgin's Bower.
In thickets along streams from the Missouri westward.

Clematis Scottii, Porter. Scott's Clematis.
In the Black Hills.

Clematis alpina tenuiloba, (Gray.) Rydberg. Alpine Clematis.
Occurs occasionally in canons in the Black Hills.

Myosurus minimus, L. Mouse-tail.
Occurs in low places in several localities in the James river valley, from the central part of the state southward; Kingsbury, Miner and Aurora counties.

Ranunculus delphinifolius, Torr. Yellow Water-crowfoot.
Ranunculus multifidus, Pursh.
　　In ponds and streams from the Missouri river eastward.

Ranunculus ovalis, Raf. Prairie Crowfoot.
Ranunculus rhomboideus, Goldie.
　　On prairies and banks of streams throughout the state.

Ranunculus abortivus, L. Kidney-leaved Crowfoot.
　　In rich woods in the Minnesota and Sioux valleys, and in the Black Hills.

Ranunculus sceleratus, L. Celery-leaved Crowfoot.
　　In the Sioux, the Minnesota and the southern part of the James valley, and in the Black Hills.

Ranunculus Pennsylvanicus, Lf. Bristly Buttercup.
　　In low, wet ground in the Minnesota, the Sioux and the James valleys, and in the Black Hills.

Ranunculus Macounii, Britton. Macoun's Buttercup.
　　A common species in the Minnesota, Sioux and James valleys, and in the Black Hills.

Ranunculus pedatifidus cardiophyllus, (Hook.) Britton.
　　Occurs in the Black Hills.

Ranunculus septentrionalis, Poir. Swamp Buttercup.
　　Common in low, wet meadows in the Minnesota and Sioux valleys.

Batrachium divaricatum, (Schrank.) Wimm. Water Crowfoot.
Ranunculus trichophyllus, Chaix.
　　Common in ponds and streams throughout the state.

Oxygraphis cymbalaria, (Pursh.) Prantl. Seaside Crowfoot.
Ranunculus cymbalaria, Pursh.
　　In sandy soil throughout the state.

Thalictrum diocium, L. Early Meadow Rue.
Occurs occasionally from the James valley eastward; Brown and Brookings counties.

Thalictrum venulosum, Trelease. Veiny Meadow Rue.
In the Black Hills.

Thalictrum occidentale, Gray. Western Meadow Rue.
Dr. Rydberg is not quite certain of his identification of this species, as he was unable to obtain fruit.

Thalictrum purpurascens, L. Purplish Meadow Rue.
In thickets and woods throughout the state.

BERBERIDACEÆ.—Barberry Family.

Berberis aquifolium, Pursh. Trailing Mahonia.
Berberis repens, Lindl.
A trailing shrub, common in canons in the Black Hills.

Caulophyllum thalictroides, (L.) Michx. Blue Cohosh.
In rich woods in the Minnesota region.

MENISPERMACEÆ.—Moonseed Family.

Menispermum Canadense, L. Canada Moonseed.
In woods and thickets along streams from the Missouri eastward.

PAPAVERACEÆ.—Poppy Family.

Argemone alba, Lestib. White Prickly Poppy.
In draws and on open plains in the foot-hills to the Black Hills, and ranging eastward nearly to the Missouri river.

Sanguinaria canadensis, L. Bloodroot.
In damp, rich woods in the Minnesota region.

Bicuculla cucullaria, (L.) Millsp. Dutchman's Breeches.
Dicentra cucullaria, DC.
In rich woods in the Minnesota and Sioux valleys.

Capnoides aureum, (Willd.) Kuntze. Golden Corydalis.
Corydalis aurea, Willd..
  In light soil in the Minnesota and Sioux valleys, and in the Black Hills. It was collected along a railroad embankment, and was possibly introduced from farther east.

Capnoides curvisiliqum, (Englm.) Kuntze. Curved-fruited Corydalis.
Corydalis curvisiliqua, Englm.
  Common in the Black Hills.

CRUCIFERÆ.—Mustard Family.

Stanleya pinnata, (Pursh.) Britton. Stanleya.
  On dry prairies in the Black Hills region.

Lepidium virginicum, L. Wild Pepper-grass.
  A weed in fields and roadsides in the Minnesota and Sioux valleys.

Lepidium apetalum, Willd. Apetalous Pepper-grass.
Lepidium intermedium, Gray.
  A very abundant weed in cultivated fields and waste places from the Missouri river eastward.

Lepidium incisum, Roth. Cut-leaved Pepper-grass.
  Occurs rarely in the Black Hills.

Thalaspi arvense, L. Field Pepper-grass.
  Sparingly introduced in the Sioux valley near Sioux Falls. Doubtless introduced from Manitoba, where it is a troublesome weed.

Sisymbrium officinale, (L.) Scop. Hedge Mustard.
  A common weed in waste places from the Missouri river eastward.

Sisymbrium altissimum, L. Tumbling Mustard.
  A common and troublesome weed in the Minnesota valley, doubtless introduced from Assinoboia.

Brassica nigra, (L.) Koch. Black Mustard.
  A common introduced weed in waste places.

Brassica arvensis, (L.) B. S. P.   Wild Mustard.
Brassica sinapistrum, Bois.
   A troublesome weed in cultivated fields throughout the state, but more abundant in the eastern part. It is much more abundant and harder to eradicate in the bottom lands.

Sinapsis alba, L.   White Mustard.
Brassica alba, Bois.
   A weed in cultivated and waste places throughout the state.

Roripa siniata, (Nutt.) A. S. Hitchcock.   Spreading Yellow-cress.
   Occasional in low places from the Missouri valley eastward.

Roripa palustris, (L.) Bess.   Marsh Water-cress.
Nasturtium palustris, DC.
   In low, wet places throughout the state.

Roripa hispida, (Desv.) Britton.   Hispid Yellow-cress.
Nasturtium hispidium, DC.
   Occasional in low places from the Missouri valley eastward.

Roripa nasturtium, (L.) Rusby.   Water-cress.
Nasturtium officinale, R. Br.
   Naturalized near Hot Springs in the Black Hills.

Roripa armoracia, (L.) A. S. Hitchcock's Horseradish.
Nasturtium armoracia, Fries.
   Sparingly introduced into the Sioux valley.

Cardamine bulbosa, (Schreb.) B. S. P.   Bulbous Cress.
Cardamine rhomboida, DC.
   In shallow water and low meadows in the Minnesota and Sioux valleys.

Cardamine rotundifolia, Michx.   American Water-cress.
   In cold, springy bogs in the Minnesota region.

Cardamine pratensis, (L.) Cuckoo-flower.

Specimens collected in the cold spring swamp near Lake Traverse in the Minnesota region, one doubtfully referred to this species.

Physaria didymocarpa, (Hook.) Gray. Double Bladder-pod.

On the dry plains east of the Black Hills.

Lesquerella spathulata, Rydberg. Low Bladder-pod.

On dry knolls in the Black Hills and the surrounding plains.

Lesquerella argentia arenosa, (Richards.) Wats. Silvery Bladder-pod.

Vesicaria arenosa, Richards.

In the Black Hills and the adjacent plains.

Bursa Bursa-pastoris, (L.) Britton. Shepard's Purse.

Capsella Bursa-pastoris, Medic.

Sparingly introduced from the Missouri eastward and in the Black Hills.

Camelina sativa, (L.) False Flax.

Occasional in flax fields and waste places in the Sioux valley and the Black Hills.

Draba Caroliniana, Walt. Carolina Whitlow-grass.

On dry, sandy knolls from the James valley westward.

Draba Caroliniana micrantha, (Mott.) Gray.

In the Black Hills.

Draba nemorasa, L. Wood Whitlow-grass.

Occasional on sandy knolls in the Sioux valley and in the Black Hills.

Draba aurea, Vahl. Golden Whitlow-grass.

In the Black Hills.

Sophia incisa, (Engelm.) Greene. Western Tansy Mustard.

Sisymbrium incisum, Engelm.

Occasional throughout the state.

Sophia Hartwegiana, (Fourn.) Greene. Hastings' Tansy Mustard.
Sisymbrium Hartwegianum, Fourn.
Common in dry soil in the Sioux valley.

Arabis hirsuta, (L.) Scop. Hairy Rock-cress.
On prairies in the Minnesota, Sioux and James valleys and in the Black Hills.

Arabis Canadensis, L. Sickle-pod.
In open woods in the Minnesota valley.

Arabis brachycarpa, (T. & G.) Britton. Purple Rock-cress.
In woods in the Minnesota valley.

Arabis glabra, (L.) Bernh. Tower Mustard.
Arabis perfoliata, L.
Occasional in the Minnesota and Sioux valleys and in the Black Hills.

Arabis Horboellii, Hormen. Horboell's Rock-cress.
In the Black Hills.

Erysimum cheiranthoides, (L.) Treachle Mustard.
In thickets in the vicinity of streams in the Minnesota and Sioux valleys and in the Black Hills.

Erysimum inconspicuum, (S. Wats.) MacM. Small Erysimum.
In the Sioux valley and the Black Hills; not common.

Erysimum Syrticolum, Sheldon. Sand Erysimum.
In the Minnesota region, near Bigstone lake.

Erysimum asperum, DC. Western Wall-flower.
On dry soil from the Missouri valley westward.

Matthiola fenestralis, Stock.
There is a single specimen in the herbarium from Spring lake, Walworth county, with no note as to the extent to which it has become naturalized.

Coringia orientalis, (L.) Dumort. Hare's-ear Mustard.
Sparingly introduced from the Missouri valley eastward, becoming troublesome in some localities.

CAPPARIDACEÆ.—Caper Family.

Cleoma serrulata, Pursh. Pink Cleome.
Cleoma integifolia, T. & G.
From the Missouri valley westward; not abundant.

Polanisia trachysperma, T. & G. Clammy Weed.
On sandy and gravelly shores throughout the state.

CRASSULACEÆ.—Orpine Family.

Sedum stenopetalum, Pursh. Western Stone Crop.
On dry, rocky knolls in the Black Hills.

Penthorum sedoides, L. Virginian Stone Crop.
Along and in streams in the Minnesota and Sioux valleys; abundant.

SAXIFRAGACEÆ.—Saxifrage Family.

Saxifraga cernua, L. Nodding Saxifrage.
Near Sylvan lake, in the Black Hills.

Heuchera hispida, Pursh. Rough Heuchera, Alum-root.
In thickets from the Missouri valley eastward, and in the Black Hills.

Heuchera parviflora, Nutt. Small Flowered Alum-root.
Near Rockford in the Black Hills.

Parnassia Caroliniana, Michx. Grass of Parnassus.
In cold swamps in the Minnesota region, and one station, near Elkton, in the Sioux valley.

Parnassia parviflora, DC. Small-flowered Grass of Parnassus.
In cold swamps in the Minnesota valley and in the Black Hills.

Tellima parviflora, Hook. Small-flowered Tellima.
In the Black Hills, rare.

GROSSULARIACEÆ.—Gooseberry Family.

Ribes gracile, Michx. Missouri Gooseberry.
Common in woods and thickets from the Missouri valley eastward.

Ribes oxycanthoides, L. Northern Gooseberry.
In the Minnesota and Sioux valleys and in the Black Hills.

Ribes lacustre, (Pers.) Poir. Swamp Gooseberry.
In the Black Hills.

Ribes Setosum, Lindl. Bristly Gooseberry.
Common in the Black Hills and along streams in the adjoining plains.

Riber floridum, L'Her. Wild Black Currant.
Very common along streams from the Missouri valley eastward.

Ribes cereum, Dougl. Squaw Currant.
Common in the "draws" and canons in the Black Hills and the adjacent plains. Doubtless extending eastward nearly to the Missouri river.

Ribes aureum, Pursh. Golden or Buffalo Currant.
In thickets and along streams from the Missouri valley westward.

ROSACEÆ.—Rose Family.

Opulaster opulifolius, (L.) Kuntze. Ninebark.
Common in the Black Hills.

Opulaster monogyna, (Torr.) Kuntze. Small-flowered Ninebark.
In the Black Hills and the adjacent plains.

Spiræa salicifolia, L.   Willow-leaved Meadowsweet.

In moist ground in the Minnesota valley, and in the Sioux valley near Sioux Falls.

Spiræa lucida, Dougl.   Corymbed Spiræa.
Spiræa betulifolia, Hook.

On banks in the Black Hills.

Luetkea cæspetosa, (Nutt.) Kuntze.   Tufted Meadowsweet.
Spiræa cæspitosa, Nutt.

In the Black Hills.

Rubus parviflorus, Nutt.   Salmon-berry.
Rubus nutkanus, Mocino.

In the Black Hills.

Rubus strigosus, Michx.   Red Raspberry.

Along streams and in rocky places throughout the state.

Rubus occidentalis, L.   Black Raspberry.

From the Missouri river eastward. In general it is not as common as the last, especially rare in the Missouri valley.

Rubus Americanus, (Pers.) Britton.   Dwarf Raspberry.
Rubus triflorus, Richards.

In cold swamps in the Minnesota region and in the Black Hills.

Fragaria Virginiana, Duchesne.   Strawberry.

In rather low ground throughout the state, but not abundant.

Fragaria Americana, (Porter.) Britton.   Wood Strawberry.   Indian Strawberry.
Fragaria Vesca Americana, Porter.

In woods in the Minnesota region, and in the Black Hills.

Potentilla arguta, Pursh.   Tall Cinquefoil.

On dry prairies throughout the state.

Potentilla nivea dissecta, Wats.   Snowy Cinquefoil.
    A rare plant, occurring in the Black Hills near Hot Springs.

Potentilla Monspeliensis, L.   Rough Cinquefoil.
    Potentilla Norvegica, L.
    In dry soils throughout the state.

Potentilla leucocarpa, Rydberg.   Diffuse Cinquefoil.
    Potentilla rivalis millegrama, S. Wats.
    In the Black Hills and the Sioux valley.

Potentilla paradoxa, Nutt.   Bushy Cinquefoil.
    Potentilla supina, Michx.
    In sandy soil throughout the state; not abundant.

Potentilla Hippiana, Lehm.   Woolly Cinquefoil.
    Occurs only in the Black Hills.

Potentilla hippiana diffusa, (Gray.) Lehm.
    Occurs with the type in the Black Hills.

Potentilla Pennsylvanica, L.   Prairie Cinquefoil.
    On prairies from the Missouri valley eastward and in the Black Hills.

Potentilla fruiticosa, L.   Shrubby Cinquefoil.
    In moist, rocky places in the Black Hills.

Potentilla anserina, L.   Silver-weed.
    In alkaline places from the Missouri river eastward.

Potentilla glandulosa, Lindl.   Glandular Cinquefoil.
    In the Black Hills.

Potentilla gracilis, Dougl.   Slender Cinquefoil.
    In the Black Hills.

Potentilla gracilis fastigiata, (Nutt.) Wats.
    Near Rockford in the Black Hills.

Potentilla concinna, Richards.   Elegant Cinquefoil.
    Near Custer in the Black Hills.

Potentilla concinna humistrata, Ryd.
    In the Black Hills, in the vicinity of Deadwood.

Geum ciliatum, Pursh. Long-plumed Purple Avens.
Geum triflorum, Pursh.
  On prairies and in thickets throughout the state.

Geum Canadensis, Jacq. White Avens.
  In woods and thickets from the Missouri river eastward.

Geum Virginianum, L. Rough Avens.
  In low ground in the Minnesota and Sioux valleys.

Geum Macrophyllum, Willd. Large-leaved Avens.
  Near Rockford in the Black Hills.

Geum strictum, Ait. Yellow Avens.
  Occasional throughout the state.

Cercocarpus parvifolius, H. & A. Small-leaved cercocarpus.
  On dry, rocky knolls in the Black Hills.

Agrominia hirsuta, (Muhl.) Becknell. Hairy Agrimonia.
Agrimonia eupatoria hirsuta, Muhl.
  In thickets in the Minnesota and James valley and in the Black Hills.

Agrimonia parviflora, Soland. Many-flowered Agrimonia.
  Occasional in woods and thickets in the Minnesota and Sioux valleys and in the Black Hills.

Rosa blanda, Ait. Meadow Rose.
  In thickets in the Minnesota and Sioux valleys.

Rosa acicularis. Prickly Rose.
  Abundant in the Black Hills.

Rosa woodesii, Lindl. Wood Rose.
  From the James valley westward, common only in the Black Hills.

POMACEÆ.—Apple Family.

Sorbus sambucifolia, (C. & S.) Roem. Western Mountain Ash.

Pyrus sambucifolia, C. &. S.
Doubtfully reported by Dr. Rydberg as occurring near Sturgis in the Black Hills.

Amelanchier Canadensis, (L.) Medic. June Berry, Service Berry.
Along the Sioux river near Sioux Falls.

Amelanchier rotundifolia, (Michx.) Roem. Round-leaved June Berry.
Amelanchier Canadensis rotundifolia, T. & G.
In woods and thickets in the Minnesota and Sioux valleys.

Amelanchier alnifolia, Nutt. Northwestern June Berry.
Amelanchier Canadensis alnifolia, T. & G.
In thickets and on dry knolls from the James valley westward; common.

Crataegus macracantha, Lodd. Long Spined Thorn-apple.
Cratægus coccinea macracantha, Dudley.
Common in thickets in the Minnesota, Sioux and James valleys, and in the Black Hills.

Crataegus mollis, (T. & G.) Scheele. Red-fruited Thorn.
Cratægus tomentora mollis, Gray.
Occasional in thickets in the Minnesota, Sioux and James valleys.

DRUPACEÆ.—Plum Family.

Prunus Americana, Marsh. Wild Plum.
Very abundant in thickets in the vicinity of lakes and streams throughout the state.

Prunus Besseyi, Bailey. Western Sand Cherry.
On bluffs and banks of streams from the James valley westward.

Prunus Pennsylvanica, L. F. Pin Cherry or Wild Red Cherry.
This species has only been reported from the Black Hills.

Prunus Virginiana, L. Choke Cherry.
   Along streams and lakes throughout the state.
Prunus demissa, (Nutt.) Walp. Western Wild Cherry.
   Along streams from the Missouri valley westward.
Prunus serotina, Ehrh. Black Cherry.
   Rare in the Minnesota and Sioux valleys.

MIMOSACEÆ.—Mimosa Family.

Acuan Illinoensis, (Michx.) Kuntze.
   Desmanthus brachylobus, Benth.
   On sandy or rocky banks of streams and lakes; occasional in the Minnesota, Sioux and James valleys.
Morongia uncinata, (Willd.) Britton. Sensitive-brier.
   Schrankia uncinata, Willd.
   In dry soil from the Missouri valley westward.

CÆSALPINACEÆ.—Senna Family.

Cercis Canadensis, L. Red-bud.
   This species is reported by Engelman to be found at the mouth of the Sioux river. It has not yet been collected there or elsewhere in the state. If it reaches our southeastern limit it must be quite rare.
Cassia chamæcrista, L. Sensitive Pea.
   In the southern part of the Sioux and James valleys, near Sioux Falls, Elk Point and Yankton.
Gleditsia triacanthos, L. Sweet Locust.
   Along the Sioux river in the extreme southeastern county, Union county.
Gymnocladus dioica, (L.) Koch. Kentucky Coffee-tree.
   Along the Missouri river in the two southeastern counties, Clay and Union.

PAPILLIONACEÆ.—Pea Family.

Sophora sericea, Nutt. Silky Sophora.
   On prairies from the Missouri valley westward.

Thermopsis rhombifolia, (Nutt.) Richards. Prairie Thermopsis.
On banks and in draws from the Missouri valley westward.

Crotalaria sagitalis, L. Rattle-box.
On prairies in the southern part of the state, east of the Missouri river; Union, Clay, Yankton and Charles Mix counties.

Lupinus sericeus, Pursh. Woolly Lupine.
In the Black Hills.

Lupinus parviflorus, Nutt. Small-flowered Lupine.
Common in the Black Hills.

Lupinus pusillus, Pursh. Low Lupine.
Common on the dry plains from the Missouri valley westward.

Melilotus alba, Desv. White Sweet Clover.
Sparingly naturalized in the Minnesota, Sioux and James valleys.

Melilotus officinalis, (L.) Lam. Sweet Yellow Clover.
Escaped from cultivation in the Sioux valley near Brookings.

Trifolium procumbens, L. Low Hop Clover.
Sparingly naturalized in the Sioux valley.

Trifolium stoloniferum, Muhl. Running Buffalo Clover.
In low prairies and woods in the Minnesota and Sioux valleys; introduced.

Trifolium Beckwithii, Brewer. Beckwith's Clover.
Very abundant around swails and in low meadows in the Minnesota and Sioux valleys. Our only native clover.

Trifolium repens, (L.) White Clover.
Sparingly introduced in the Sioux valley and in the Black Hills.

Trifolium pratense, L. Red Clover.
Sparingly naturalized in the Minnesota and Sioux valleys.

Lotus Americanus, (Nutt.) Bisch. Prairie Bird's-foot Trefoil.
Hosackia purshiana, Benth.
On low, sandy land, mostly in the vicinity of streams, locally very abundant; it is the most nutritious of all our native forage crops. In cultivation, however, its growth is uncertain. The name of "Dakota Vetch" has been applied to it by J. G. Smith.

Psoralea tenuiflora, Pursh. Few-flowered Psoralea.
Common in dry soils from the Missouri river valley westward.

Psoralea digitata, Nutt. Digitate Psoralea.
Occasional in dry soils from the Missouri river westward.

Psoralea argophylla, Pursh. Silver-leaf Psoralea.
Very abundant throughout the state.

Psoralea cuspidata, Pursh. Large-bracted Psoralea.
Common on dry hills and banks from the Missouri valley westward.

Psoralea esculenta, Pursh. Prairie Turnip, "Indian Turnip."
On sandy knolls throughout the state; common.

Psoralea lanceolata, Pursh. Lance-leaved Psoralea.
In the southern part of the Sioux valley; uncommon.

Amorpha fruticosa, L. False Indigo. Lead Plant.
Bordering streams and lakes throughout the state.

Amorpha nana, Nutt. Fragrant False Indigo.
Amorpha microphylla, Pursh.
On banks and prairies from the James valley westward to the Black Hills; also reported from Sioux Falls in the Sioux valley; more abundant from the Missouri river westward.

Amorpha canescens, Pursh. Shoe-strings. Lead-Plant.
  Very abundant on prairies throughout the state.
Parosela enneandra, (Nutt.) Britton. Slender Parosela.
  Dalea laxiflora, Pursh.
  Common from the Missouri valley to the Black Hills.
Parosela dalea, (L.) Britton. Pink Parosela.
  Dalea alopecuroides, Willd.
  In sandy soils from the Missouri valley eastward; occasional in the Minnesota and Sioux valleys; common in the Missouri valley.
Parosela aurea, (Nutt.) Britton. Golden Parosela.
  Dalea aurea, (Nutt.)
  On bluffs and loose soils from the Missouri valley westward.
Kuhnistera candida, (Willd.) Kuntze. White Prairie Clover.
  Petalstemon candidus, Michx.
  Very common on prairies throughout the state.
Kuhnistera compacta, (Spreng.) Kuntze. Dense-flowered Prairie Clover.
  Petalstemon compacta, Swezey.
  Occasional on dry plains from the Missouri valley westward.
Kuhnistera purpurea, (Vent.) MacM. Violet Prairie Clover.
  Petalstemon violaceus, Michx.
  Common throughout the state; especially abundant in sandy or gravelly soils.
Kuhnistera villosa, (Nutt.) Kuntze. Hairy Prairie Clover.
  Petalstemon villosus, Nutt.
  In the Bad Lands, east of the Black Hills.
Astragalus crassicarpus, Nutt. Buffalo Pea, Ground Plum.
  Astragalus caryocarpus, Ker.
  Very common on prairies throughout the state.

Astragalus Mexicanus, DC.   Larger Ground Plum.
   In the Sioux valley, near Brookings.
Astragalus Plattensis, Nutt.   Platte Milk Vetch.
   In the Sioux valley near Sioux Falls, and in the Black Hills.
Astragalus Carolinianus, L.   Carolina Milk Vetch.
   Astragalus Canadensis, L.
   In sandy soil and waste places in the Minnesota and Sioux valleys and in the Black Hills.
Astragalus adsurgens, Pall.   Ascending Milk Vetch.
   On prairies throughout the state.
Astragalus hypoglottis, L.   Cock's-head.
   Abundant from the Missouri valley eastward, and in the Black Hills.
Astragalus Drummondii, Dougl.   Drummond's Milk Vetch.
   In the Black Hills, near Hot Springs.
Astragalus racemosus, Pursh.   Racemose Milk Vetch.
   Abundant in heavy soils from the Missouri river westward; occasional in the James valley.
Astragalus bisulcatus, (Hook.) Gray.   Grooved Milk Vetch.
   In the foot-hills of the Black Hills, and in the adjoining plains.
Astragalus lotiflorus, Hook.   Low Milk Vetch.
   In the Black Hills and on the adjoining plains.
Astragalus Missouriensis, Nutt.   Missouri Milk Vetch.
   From the Missouri valley westward; common on dry banks and hills.
Astragalus gracilis, Nutt.   Slender Milk Vetch.
   In dry ground in the Minnesota region, and near Hot Springs in the Black Hills.
Astragalus microlobus, Gray.   Notched Milk Vetch.
   Common in the Black Hills.

Astragalus flexuosus, (Hook.) Dougl. Flexile Milk Vetch.
  On dry soils throughout the state.
Astragalus aboriginum, Richards. Indian Milk Vetch.
  Near Deadwood, in the Black Hills.
Astragalus aboriginum glaberiusculus, (Hook.) Rydberg. Indian Milk Vetch.
  In the Black Hills.
Astragalus alpinus, L. Alpine Milk Vetch.
  In woods in the Black Hills.
Astragalus convallarius, Greene.
  Near Bull Sprigs in the Black Hills.
Phaca Americana, (Hook.) Rydberg. Arctic Milk Vetch.
  Phaca frigida Americana, Hook.
  On low ground in thickets in the Black Hills and the adjacent plains.
Homalobus tenellus, (Pursh.) Britton. Loose-flowered Milk Vetch.
  Astragalus tenellus, Pursh.
Homalobus montanus, (Nutt.) Britton. Prickly Milk Vetch.
  Astragalus Kentrophyta, A. Gray.
  In the Bad Land region, east of the Black Hills.
Homalobus caespitosus, Nutt. Tufted Milk Vetch.
  Astragalus cæspitosus, A. Gray.
  In dry soils in the Black Hills, and in the adjacent plains region.
Orophaca caespitosa, (Nutt.) Britton. Sessile-flowered Milk Vetch.
  Astragalus cæspitosa, Nutt.
  In dry soils from the Missouri river westward.
Spiesia Lamberti, (Pursh.) Kuntze. Loco-weed.
  Oxytropis Lamberti, Pursh.
  Abundant on high ground throughout the state.

Spiesia Lamberti sericea, (Nutt.) Rydberg.
Oxytropus sericea, Nutt.
   Throughout the state, but more abundant from the Missouri valley westward.
Spiesia viscida, (Nutt.) Kuntze.   Viscid Loco-weed.
   Common near Custer in the Black Hills.
Glycyrrhiza lepidota, Pursh.   Wild Licorice.   Buffalo Burr.
   Very abundant in low ground throughout the state.
Hedysarum Americanum, (Michx.) Britton.   Hedysarum.
   In the Black Hills near Rockford.
Meibomia grandiflora, (Walt.) Kuntze.   Broad-leaved Tick-trefoil.
Desmodium acuminatum, Michx.
   In woods in the Minnesota and Sioux valleys and up the Missouri valley to Yankton.
Meibomia Illinoensis, (A. Gray.) Kuntze.   Illinois Tick-trefoil.
Desmodium Illinoense, A. Gray.
   Common in thickets and low prairies in the Minnesota valley.
Lespedeza capitata, Michx.   Round-headed Bush Clover.
   In low, sandy soil in the southern part of the Sioux valley; Sioux Falls and Elk Point; up the Missouri to Yankton.
Vicia Americana, Muhl.   American Vetch.
   In woods and thickets in the Minnesota and Sioux valleys and the Black Hills.
Vicia Americana truncata, (Nutt.) Brewer.
Vicia truncata, Nutt.
   In or near low ground throughout the state.
Vicia linearis, (Nutt.) Greene.   Narrow-leaved Vetch.
Vicia Americana linearis, S. Wats.
   Common on dry prairies throughout the state.

Vicia sativa, L. Vetch. Tare.
Introduced into waste or cultivated fields in the Sioux valley.

Lathyrus venosus, Muhl. Veiny Pea, Wild Pea.
On shady banks of streams and lakes in the Minnesota and Sioux valleys.

Lathyrus palustris, L. Marsh Vetchling.
In low, moist ground from the Missouri river eastward.

Lathyrus ornatus, Nutt. Showy Vetchling.
On banks and bluffs from the Missouri valley westward.

Lathyrus ochroleucus, Hook. Cream-colored Vetchling.
Abundant in the Black Hills.

Falcata comosa, (L.) Kuntze. Hog Peanut.
Amphicarpa monoica, Ell.
Near Rapid City in the Black Hills.

Falcata Pitcheri, (T. & G.) Kuntze. Pitcher's Hog Peanut.
Amphicarpa Pitcheri, T. & G.
In rich woods in the Minnesota valley.

Apois apois, (L.) McM. Groundnut.
Apois tuberosa, Moench.
In woods and thickets in the Minnesota and Sioux valleys.

Strophostyles helvola, (L.) Britton. Trailing Wild Bean.
Strophostyles angulosa, Ell.
On shady banks of streams and lakes from the Missouri valley eastward; not common.

Strophostyles pauciflora, (Benth.) S. Wats. Small Wild Bean.
On banks of streams and lakes in the Minnesota, Sioux and James valleys.

GERANIACEÆ.—Geranium Family.

Geranium columbinum, L. Long-stalked Crane's Bill.
 In the southern part of the Missouri valley; Charles Mix county.

Geranium Carolinianum, L. Carolina Crane's Bill.
 In the Black Hills and the southern part of the Missouri valley.

Geranium dissectum, L. Cut-leaved Crane's Bill.
 A native of Europe; near Rapid City in the Black Hills; rare.

Geranium Richardsonii, Fisch. & Meyers. Richardson's Crane's Bill.
 The common species in the Black Hills.

Geranium viscosissimum, Fisch. & Meyers. Viscid Crane's Bill.
 Occasional in the Black Hills.

OXALIDACEÆ.—Wood-sorrel Family.

Oxalis violacea, L. Violet Wood-sorrel.
 In low places in thickets and in open prairies in the Minnesota, Sioux and James valleys.

Oxalis stricta, L. Upright Yellow Wood-sorrel.
 In thickets and on prairies from the Missouri valley eastward and in the Black Hills.

LINACEÆ.—Flax Family.

Linum Lewisii, Pursh. Lewis' Wild Flax.
 In the Black Hills.

Linum sulcatum, Riddell. Grooved Yellow Flax.
 In dry soil in the Minnesota, Sioux and James valleys.

Linum rigidum, Pursh. Large-flowered Yellow Flax.
 Common on dry prairies throughout the state.

RUTACEÆ.—Rue Family.

Xanthoxylum Americanum, Mill. Prickly Ash.
In woods along streams and bordering lakes from the Missouri valley eastward.

POLYGALACEÆ.—Milkwort Family.

Polygala verticillata, L. Whorled Milkwort.
On bluffs and prairies throughout the state.

Polygala senega latifolia, T. & G.
Common in the northern part of the Black Hills.

Polygala alba, Nutt. White Milkwort.
Common from the Missouri valley eastward and in the Black Hills.

EUPHORBIACEÆ.—Spurge Family.

Croton Texensis, (Klotzsch.) Muell. Texas Croton.
Common in dry, sandy soils from the Missouri valley westward.

Euphorbia petaloidea, Engelm. White-flowered Spurge.
In sandy soil from the Missouri valley westward to the Black Hills.

Euphorbia Geyeri, Engelm. & Gray. Geyer's Spurge.
In dry soils from the Missouri valley westward.

Euphorbia serpyllifolia, Pers. Thyme-leaved Spurge.
Common in the Minnesota, Sioux, and James valleys.

Euphorbia serpens, H. B. K. Round-leaved Spurge.
In the southern Missouri valley; Charles Mix county.

Euphorbia Glyptosperma, Engelm. Ridge-seeded Spurge.
In dry soils throughout the state; the most common species.

Euphorbia maculata, L. Milk Purslane.
In the Sioux and James valleys.

Euphorbia stictospora, Engelm. Narrow-seeded Spurge.
   Common in dry soils from the Missouri valley westward.
Euphorbia nutans, Lag.   Upright Spotted Spurge.
   Euphorbia preslii, Guss.
   In the Sioux, James and Missouri valleys, especially the eastern part.
Euphorbia hexagona, Nutt.   Angled Spurge.
   In dry soils from the Missouri valley westward.
Euphorbia marginata, Pursh.   White Margined Spurge.
   Along the Missouri river in the southern part of the state and thence westward.
Euphorbia dentata, Michx.   Toothed Spurge.
   In the southern part of the Missouri valley, Charles Mix county, and in the Black Hills.
Euphorbia heterophylla, L.   Various-leaved Spurge.
   In the southern part of the Sioux and Missouri valleys; near Sioux Falls, Yankton and Running Water.
Euphorbia dictyosperma, F. & M.   Reticulate-seeded Spurge.
   In the James valley near Aberdeen, and from thence westward.
Euphorbia robusta, (Engelm.) Small.   Rocky Mountain Spurge.
   Euphorbia mountain robusta, Engelm.
   Along the Cheyenne river and in the Black Hills; probably ranging from the Missouri valley westward.

CALLITRICHACEÆ.—Water Starwort Family.

Callitriche palustris, L.   Water Fennel.
   Common in clear, cold streams throughout the state.
Callitriche bifida, (L.) Morong.   Autumnal Starwort.
   Callitriche autumnalis, L.
   In running water in the Minnesota and Sioux valleys.

ANACARDIACEÆ.—Sumac Family.

Rhus hirta, (L.) Sudw. Staghorn Sumac.
Rhus typhina, L.
Collected but once in the southern part of the Missouri valley, in Charles Mix county.

Rhus glabra, L. Smooth Sumac.
Common on bluffs and banks of streams from the Missouri valley eastward, and in the Black Hills.

Rhus trilobata, Nutt. Skunkbush.
On dry blufts and hills from the Missouri valley westward, common; many specimens have very pubescent leaflets.

Rhus radicans, L. Poison Oak. Poison Ivy.
In woods and thickets throughout the state.

CELASTRACEÆ.—Staff-tree Family.

Euonymus atropurpureous, Jacq. Burning Bush. Wahoo.
Occasional in woods in the Sioux valley, and up the Missouri river into Charles Mix county.

Celastrus scandens, L. Bittersweet.
Common in woods and thickets throughout the state.

ACERACEÆ.—Maple Family.

Acer saccharinum, L. Silver Maple. Soft Maple.
Acer dasycarpum, Ehrh.
Extends up the Sioux river to Flandreau and up the Missouri river to the western edge of Union county—the southeastern county.

Acer saccharum, Marsh. Sugar Maple. Rock Maple.
Acer saccharum, Wang.
Abundant in coulees and ravines, containing a clear spring creek, at the headwaters of the Little Minnesota river. There are some ten coulees extending from one

to three miles back into the coteaus in which the sugar maple is abundant. Many of the trees are sixty to seventy feet high and eight to ten feet in circumference. As soon as the creek emerges from the hills into the open prairie all the trees except the white elm, box elder and two willows disappear. The coulees are narrow and deep, and the banks are lined with springs which furnish a constant supply of water. The southernmost coulee in which maples were found is one-half mile south of Sisseton Agency.

Acer Negundo, L. Box Elder. Ash-leaved Maple.
Abundant along lakes and streams throughout the state.

BALSAMINACEÆ.—Jewel-weed Family.

Impatiens biflora, Walt. Spotted Touch-me-not.
Impatiens fulva.
In damp woods in the Minnesota and Sioux valleys.

Impatiens aurea, Muhl. Pale Touch-me-not.
Impatiens pallida, Nutt.
In moist woods in the Minnesota valley.

RHAMNACEÆ.—Buckthorn Family.

Ceanothus velutinus, Dougl. Velvety Redroot.
In the Black Hills near Lead City.

Ceanothus Fendleri, Gray. Fendler's Redroot.
Common in limestone districts in the Black Hills.

VITACEÆ.—Grape Family.

Vitis Vulpina, L. Riverside Grape.
Vitis riparia, Michx.
Common along streams and bordering lakes from the Missouri river eastward.

Parthenocissus quinquefolia, (L.) Planch. False Virginia Creeper. Woodbine.

Ampelopsis quinquefolia, Michx.
Common in same habitat as the last, the Minnesota and Sioux valleys, and in the Black Hills.

TILIACEÆ.—Linden Family.

Tilia Americana, L. Basswood. American Linden.
Along streams and bordering lakes in the Minnesota and Sioux valleys; up the Missouri valley to Running Water.

MALVACEÆ.—Mallow Family.

Malva Sylvestris, L. High Mallow.
Sparingly introduced into waste places in the Sioux valley near Brookings.

Malva rotundifolia, L. Cheese's Running Mallow.
A naturalized weed in waste places in the Sioux valley and southern Missouri valley.

Malva verticillata crispa, L. Whorled Mallow.
In waste places in the Sioux valley.

Malvastrum coccineum, (Pursh.) A. Gray. Red Mallow.
Common on dry prairies from the James valley westward.

Abutilon abutilon, (L.) Rusby. Indian Mallow.
Abutilon avicennæ, Gaertn.
An introduced weed in the southeastern part of the state.

Hibiscus trionum, L. Venice Mallow. Flower-of-an-hour.
In waste places along the Sioux valley in Union county, and up the Missouri valley to Chamberlain, Brule county.

HYPERICACEÆ.—St. Johnswort Family.

Hypericum Canadense, L. Canadian St. Johnswort.
In low ground in the Minnesota valley, and in the Black Hills.

ELATINACEÆ.—Waterwort Family.

Elatine triandra, Schk.   Long-stemmed Waterwort.
In mud on margins of ponds and rivers in the Minnesota valley; in the Missouri valley in Walworth county, and in White river in the plains region.

CISTACEÆ.—Rock Rose Family.

Helianthemum majus, (L.) B. S. P.   Hoary Frostweed.
On roadsides east of Custer in the Black Hills.

VIOLACEÆ.—Violet Family.

Viola pedatifida, Don.   Prairie Violet.
On prairies in the Minnesota, Sioux and James valleys, and in the Black Hills.

Viola obliqua, Hill.   Meadow Violet.
Viola cucullata, Ait.
Mostly along streams or in the shady places from the Missouri river eastward, and in the Black Hills.

Viola pedata, L.   Bird's-foot Violet.
On prairies in the Minnesota valley near Bigstone Lake.

Viola palustris, L.   Marsh Violet.
In moist soil near Sylvan Lake in the Black Hills.

Viola blanda, Willd.   Sweet White Violet.
In damp ground in the Black Hills.

Viola Nuttallii, Pursh.   Nuttall's Violet.
On prairies from the James valley westward throughout the state.

Viola pubescens, Ait.   Hairy Yellow Violet.
In rich woods in Union county in the Sioux valley, and in the Black Hills. Most of the specimens referred to this species belong to the next.

Viola scabriuscula, (T. & G.) Schwein.   Smoothish Yellow Violet.
Viola pubescens scabriuscula, T. & G.

In rich woods in the Minnesota and Sioux valleys, and in the Black Hills.

Viola Canadensis, L. Canadian Violet.
Abundant in rich woods in the Minnesota and Sioux valleys, and in the Black Hills.

Viola arenaria, Dc. Sand Violet.
In dry soils in the Black Hills, near Rapid City.

Viola canina adunca, Gray.
In the Black Hills.

LOASACEÆ.—Loasa Family.

Mentzelia oligosperma, Nutt. Few-seeded Mentzelia.
Occasional in the Black Hills; more common in the adjacent plains.

Mentzelia nuda, (Pursh.) T. & G. Bractless Mentzelia.
A rare plant, reported only for the Black Hills.

Mentzelia decapetala, (Pursh.) Urban & Gilg. Showy Mentzelia.
Mentzelia ornata, T. & G.
Common in dry soils from the Missouri westward.

CACTACEÆ.—Cactus Family.

Opuntia humifusa, Raf. Western Prickly Pear.
Opuntia Rafinesquii, Engelm.
In the Black Hills and the adjacent plains region.

Opuntia fragilis, (Nutt.) Haw. Brittle Opuntia.
In the Black Hills; rare.

Cactus Missouriensis, (Sweet.) Kuntze. Missouri Cactus.
Mamillaria Missouriensis, Sweet.
On bluffs, knolls and plains from the Missouri river westward.

Cactus viviparus, Nutt. Purple Cactus.
Mamillaria vivipara, Haw.
From the Missouri valley westward.

ELAEAGNACEÆ.—Oleaster Family.

Elaeagnus argentea, Pursh. Silver Berry.
In limestone districts in the Black Hills.

Lepargyræa Canadensis, (L.) Greene. Canadian Buffalo Berry.
Shepherdia Canadensis, Nutt.
Occasional from the Missouri river to the Black Hills.

Lepargyræa argentea, (Nutt.) Greene. Buffalo Berry.
Shepherdia argentea, Nutt.
Common on bluffs and in thickets from the Missouri valley westward; occasional on bluffs in coulees in the Minnesota valley.

LYTHRACEÆ.—Loosestrife Family.

Ammania coccinea, Rottb. Long-leaved Ammannia.
In swampy ground from the Missouri valley eastward.

Lythrum alatum, Pursh. Wing-angled Loosestrife.
Occasional in low ground from the Missouri valley eastward.

ONAGRACEÆ.—Evening Primrose Family.

Chamænerion angustifolium, (L.) Scop. Fireweed. Great Willow Herb.
Epilobium angustifolium, L.
In open woods in the Minnesota valley and in the Black Hills.

Epilobium lineare, Muhl. Linear-leaved Willow Herb.
In swamps in the Minnesota and Sioux valleys, and in the Black Hills.

Epilobium paniculatum, Nutt. Panicled Willow Herb.
Occurs in the Black Hills.

Epilobium coloratum, Muhl. Purple-leaved Willow Herb.
In low ground in the Minnesota valley.

Epilobium adenocaulon, Haussk. Northern Willow Herb.
In moist ground throughout the state.

Epilobium Hornemanni, Rerchenb. Hornemann's Willow Herb.
Near Sips Springs in the Black Hills.

Epilobium Drummondii, Kansk. Drummond's Willow Herb.
In the limestone district in the Black Hills.

Onagra biennis, (L.) Scop. Evening Primrose.
Œnothera biennis, L.
Common throughout the state.

Œnothera laciniata, Hill. Sinuate-leaved Evening Primrose.
Œnothera sinuata, L.
In the Black Hills; rare.

Anogra albicaulis, (Pursh.) Britton. Prairie Evening Primrose.
Œnothera albicaulis, Pursh.
Occasional in low ground throughout the state.

Anogra coronopifolia, (T. &. G.) Britton. Cut-leaved Evening Primrose.
Œnothera coronopifolia, T. & G.
In the Black Hills and on the adjacent plains.

Anogra pallida leptophylla, (Nutt.)
Œnothera pallida leptophylla, (Nutt.) T. & G.
Œnothera leptophylla, Nutt.
Occasional in the Black Hills,

Pachylophus cæspitosa, (Nutt.) Raimann. Scapose Primrose.
Œnothera cæspitosa, (Nutt.)
In "gumbo soils" from the Missouri valley westward; common.

Meriolix serrulata, (Nutt.) Walp. Tooth-leaved Primrose.

Œnothera serrulata, Nutt.
   In dry soils throughout the state.
Gaura parviflora, Dougl. Small-flowered Gaura.
   In dry soil from the Missouri valley westward.
Gaura coccinea, Pursh. Scarlet Gaura.
   In dry prairies throughout the state; more common in the James and Missouri valleys.
Gayophyton ramosissimum, T. & G. Bushy Gayophyton.
   On dry knolls in the Black Hills and the adjacent plains.
Circæa Lutetiana, L. Enchanter's Nightshade.
   In damp woods in the Minnesota valley and the Black Hills.
Circæa alpina, L. Alpine Enchanter's Nightshade.
   In moist woods in the Black Hills.

HALORAGIDACEÆ.—Water Millfoil Family.

Hippuris vulgaris, L. Bottle Brush.
   In shallow water in the Sioux valley and locally in the Missouri valley; Faulkton.
Myriophyllum spicatum, L. Spiked Water Millfoil.
   Common in lakes and slow streams in the Minnesota, Sioux and James valleys.
Myriophyllum verticillatum, L. Whorled Water Millfoil.
   In the Minnesota and Sioux valleys; not common.
Myriophyllum heterophyllum, Michx. Various-leaved Water Millfoil.
   In slow running water in the Sioux valley; rare.

ARALIACEÆ.—Ginseng Family.

Aralia racemosa, L. American Spikenard.
   In rich woods in the Minnesota valley.

Aralia nudicaulis, L. Wild Sarsaparilla.
On shady banks in the Minnesota valley and in the Black Hills.

UMBELLIFERÆ.—Carrot Family, Umbelworts.

Daucus carrota, L. Wild Carrot.
Sparingly naturalized in the Minnnesota and Sioux valleys.

Heracleum lanatum, Michx. Cow Parsnip.
In woods in the Minnesota and Sioux valleys, and in the Black Hills.

Pastinaca sativa, L. Wild Parsnip.
Escaped near Rapid City. in the Black Hills.

Peucedanum nudicaule, (Pursh.) Nutt. White-flowered Parsley.
In dry soil from the Missouri valley eastward, common.

Peucedanum foeniculaceum, Nutt. Fennel-leaved Parsley.
In the James and Missouri valleys and along the White river; not common.

Peucedanum villosum, Nutt. Hairy Parsley.
In dry, clay, soils from the James valley westward; common.

Cymopterus acaulis, (Pursh.) Rydberg. Plain Cymopterus.
Cymopterus glomeratus, Raf.
On dry knolls from the Missouri valley westward.

Cymopterus montanus, T. & G. Mountain Cymopterus.
In the Black Hills and the adjacent plains.

Eryngium aquaticum, L. Button Snakeroot.
In the Sioux valley near Sioux Falls; rare.

Sanicula Marylandica, L. Black Snakeroot.
In woods in the Minnesota and Sioux valleys; common.

Sanicula canadensis, L.   Short-styled Snakeroot.
Sanicula Marylandica canadensis, Torr.
   In the Minnesota and Sioux valleys and in the Black Hills.

Musineon divaricatum, (Pursh.) Nutt.   Leafy Musineon.
   In "gumbo soils" from the Missouri valley westward; common.

Musineon tenuifolium, Nutt.   Scapose Musineon.
   In the Black Hills; common.

Musineon trachysperma, Nutt.
   Near Hermosa in the Black Hills.

Washingtonia Claytoni, (Michx.) Britton.   Wooly Sweet Cicely.
Osmorrhiza brevistylis, DC.
   In the Minnesota and the southern part of the Sioux valleys (Union Co.); rare.

Washingtonia longistylis, (Tott.) Britton.   Smoother Sweet Cicely.
Osmorrhiza longistylis, Torr.
   In rich woods in the Minnesota and Sioux valleys and in the Black Hills.

Washingtonia nuda, (Torr.)   Western Sweet Cicely.
Osmorrhiza nuda, Torr.
   In the Black Hills, not common.

Sium cicutæfolium, Gmel.   Hemlock Water Parsnip.
   In swamps from the Missouri valley eastward.

Zizia aurea, (L.) Koch.   Golden Meadow Parsnip.
   Abundant in low prairies from the Missouri valley eastward.

Zizia cordata, (Walt.) DC.   Heart-leaved Alexander.
   On low prairies in the Minnesota and Sioux valleys and in the Black Hills. Less common than the last.

Carum carui, L. Caraway.
>Sparingly introduced in the Sioux valley and in the Black Hills.

Carum Gairdneri, (Nutt.) Benth. & Hook. Gairdner's Caraway.
>Reported by Dr. Gray for the Black Hills.

Cicuta maculata, L. Water Hemlock.
>In swamps and streams from the Missouri valley eastward, and in the Black Hills.

Cicuta bulbifera, L. Bulb-bearing Water Hemlock.
>In a cold spring swamp near Elkton in the extreme eastern part of the Sioux valley.

Deringia canadensis, (L.) Kuntze. Honewort.
>*Cryptotænia canadensis*, L.
>In rich woods in the Minnesota and Sioux valleys.

Berula erecta, (Huds.) Coville. Cut-leaved Water Parsnip.
>*Berula augustifolia*, Mert. & Kock.
>In swamps in the Minnesota valley and in the Black Hills.

Bupleurum rotundifolium, L. Modesty.
>Introduced into the Sioux valley in grass seed.

CORNACEÆ.—Dogwood Family.

Cornus canadensis, L. Dwarf Cornel.
>In damp woods in the Black Hills.

Cornus Baileyi, Coult. & Evans. Bailey's Dogwood.
>In the Black Hills.

Cornus asperifolia, Michx. Rough-leaved Dogwood.
>Along the Missouri river and in the southern part of the state, Union, Clay and Yankton counties.

Cornus stolonifera, Michx. Red Osier.
>In thickets along streams and bordering lakes throughout the state.

Cornus amomum, Mill.   Silky Cornel.
Cornus sericea, L.
>Along the Missouri river in Yankton and Charles Mix counties.

PYROLACEÆ.—Wintergreen Family.

Pyrola rotundifolia, L.   Round-leaved Wintergreen.
>In woods in the Black Hills, Lead City.

Pyrola Elliptica, Nutt.   Shin-leaf.
>On shady hillsides in the Black Hills.

Pyrola chlorantha, Sw.   Greenish-flowered Wintergreen.
>In woods in the Black Hills.

Pyrola secunda, L.   One-sided Wintergreen.
>In deep shady ravines in the Minnesota valley and in the Black Hills.

Pyrola rotundifolia bracteata, (Nutt.) Gray.
>In a cold bog near Sylvan lake in the Black Hills.

MONOTROPACEÆ.—Indian Pipe Family.

Pterospora andromedea, Nutt.   Pine Drops.
>In woods in the Black Hills; Custer and Rapid City.

Monotropa uniflora, L.   Indian Pipe.
>In deep, wooded ravines in the Minnesota valley; Roberts county.

ERICACEÆ.—Heath Family.

Arctostaphylos uva-ursi, Spreng.   Red Bearberry.
>On dry knolls in the Black Hills, and in the adjacent plains; Custer, and the Bad Lands.

VACCINEACEÆ.—Huckleberry Family.

Vaccinium myrtillus microphyllum, Hook.   Bilberry, Whortleberry.
>In the Black Hills.

PRIMULACEÆ.—Primrose Family.

Androsace occidentalis, Pursh. Androsace.
In dry soils from the Missouri valley eastward; common.

Androsace septentrionalis, L. Mountain Androsace.
In the Black Hills, Custer.

Androsace septentrionalis subulifera, Gray.
In the foothills of the Black Hills; near Rapid City.

Steironema ciliaum, (L.) Raf. Fringed Loosestrife.
Lysimachia ciliata, L.
In damp thickets from the Missouri valley eastward, and in the Black Hills; common.

Naumbergia thrysiflora, (L.) Duby. Tufted Loosestrife.
Lysimachia thyrsiflora, L.
In cold spring bogs in the Minnesota valley near Elkton in the Sioux valley, and in the Black Hills.

Centunculus minimus, L. Chaffweed.
In the upper Missouri valley, (Walworth and Potter counties) and the Black Hills.

Dodecatheon media, L. Shooting Star, American Cowslip.
On moist banks in the Black Hills; Rapid City, and Custer.

OLEACEÆ.—Olive Family.

Fraxinus lanceolata, Borck. Green Ash.
Fraxinus viridus, Michx. F.
Abundant along streams and bordering lakes throughout the state.

Fraxinus Pennsylvanicus, Marsh. Red Ash.
Fraxinus pubescens, Lam.
With the last throughout the state and rather more common.

Fraxinus Americana, L. White Ash.

This has been repeatedly reported for this state, but the most authentic account of it in the state is the following from Prof. Williams' note book: "Twigs sent by Mr. Jones from Sioux Falls to Dr. Trelease were pronounced to be this species." If it occurs in the Minnesota or Sioux regions it is exceedingly rare, as for two summers special effort has been made to detect it.

GENTIANACEÆ.—Gentian Family.

Gentiana detonsa, Rottb. Fringed Gentian.
Gentiana serrata, Gunner.

In cold spring bogs in the Minnesota valley; one station in the Sioux valley; Elkton.

Gentiana acuta, Michx. Northern Gentian.
Gentiana amarella acuta, Herder.

Near Custer in the Black Hills.

Gentiana puberula, Michx. Downy Gentian.

Common on prairies in the Minnesota and Sioux valleys.

Gentiana Andrewsii, Griseb. Closed Gentian.

In low places in the Minnesota and Sioux valleys; common.

Frasera speciosa, Dougl. Showy Frasera.

On dry, barren knolls in the Black Hills.

Tetragonanthus deflexus, (J. E. Smith.) Kuntze. Spurred Gentian.
Swertia deflexa, J. E. Smith.

In woods in the Black Hills.

MENYANTHACEÆ.—Buckbean Family.

Menyanthes trifoliata, L. Buckbean. Bogbean.

In a cold spring bog near Elkton in the Sioux valley.

APOCYNACEÆ.—Dogbane Family.

Apocynum androsæmifolium, L. Spreading Dogbane.
In thickets in the Minnesota and Sioux valleys and in the Black Hills.

Apocynum cannabinum, L. Indian Hemp.
On dry banks from the Missouri valley eastward.

ASCLEPIADACEÆ.—Milkweed Family.

Asclepias tuberosa, L. Pleurisy Root. Butterfly Weed.
In thickets in the southern part of the Sioux valley; Union county.

Asclepias incarnata, L. Swamp Milkweed.
In swamps from the Missouri valley eastward.

Asclepias syriaca, L. Common Milkweed.
Asclepias Cornuti, Dec.
On prairies and in thickets in the Minnesota and Sioux valleys.

Asclepias speciosa, Torr. Showy Milkweed.
In low ground in the Minnesota and Sioux valleys.

Asclepias ovalifolia, Dec. Oval-leaved Milkweed.
On prairies in the Minnesota and Sioux valleys, and in the Black Hills; rare.

Asclepias verticillata, L. Whorled Milkweed.
Common on banks and prairies from the Missouri valley eastward.

Asclepias pumila, (Gray.) Vail. Low Milkweed.
Asclepias verticillata pumila, Gray.
Common on the dry plains from the Missouri valley westward, seems to replace the last.

Acerates viridiflora, (Raf.) Eaton. Green Milkweed.
Occasional in sandy soils throughout the state.

Acerates angustifolia, (Nutt.) Dec. Narrow-leaved Milkweed.
In the Black Hills and the adjacent plains.

Acerates lanuginosa, (Nutt.) Dec. Woolly Milkweed.
  On prairies in the Sioux valley; rare.

CONVOLVULACEÆ.—Morning Glory Family.

Evolvulus pilosus, Nutt. Evolvulus. In dry plains west of the Missouri river and in the Black Hills.

Quamoclit coccinea hederifolia. Small Red Morning Glory.
  Introduced in gardens near Brookings.

Ipomœa leptophylla, Torr. Bush Morning Glory.
  In dry soils from the Missouri river to the Black Hills.

Ipomœa hederaceæ, Jacq. Ivy-leaved Morning Glory.
  A bad weed in fields in the Sioux valley near Brookings.

Convolvulus sepium, L. Hedge Bindweed.
  In thickets from the Missouri valley eastward and in the Black Hills.

Convolvulus repens, L. Trailing Bindweed.
  In cultivated and waste fields from the Missouri valley eastward.

CUSCUTACEÆ.—Dodder Family.

Cuscuta Epithymum, Murr. Clover Dodder.
  On alfalfa in the Sioux valley in Brookings and Clark counties, and in the Black Hills. An introduced parasite, doing considerable damage to alfalfa wherever it occurs in quantity.

Cuscuta arvensis, Beyrich. Field Dodder.
  On various large herbs in the Minnesota valley and in the Black Hills.

Cuscuta coryli, Engelm. Hazel Dodder.
  Cuscuta inflexa, Engelm.
  On large herbs throughout the state.

Cuscuta Gronovii, Willd. Gronovi's Dodder.
  In the Sioux and Minnesota valleys; common.

Cuscuta paradoxa, Raf. Glomerata Dodder.
Cuscuta glomerata, Choisy.
   On composites in the Minnesota and Sioux valleys.

POLEMONIACEÆ.—Phlox Family.

Phlox pilosa, L. Downy Phlox.
   On prairies in the Minnesota and Sioux valleys.

Phlox Kelseyii, Britton. Kelsey's Phlox.
   In the Black Hills.

Phlox Douglassii, Hook. Douglass' Phlox.
   Common in the Black Hills.

Phlox Douglassii andicola, Britton.
   On dry table-lands in the Black Hills.

Gilia spicata capitata, Gray. Capitate Gilia.
   On dry knolls in the Black Hills.

Collomia linearis, Nutt. Narrow-leaved Collomia.
Gilia linearis, Gray.
   On dry soils from the Missouri valley westward; found also in one station in the Sioux valley.

HYDROPHYLLACEÆ.—Waterleaf Family.

Hydrophyllum Virginicum, L. Virginian Waterleaf.
   Common in woods in the Minnesota valley; occasional in the Sioux valley.

Macrocalyx Nyctalea, (L.) Kuntze. Nyctalea.
Ellisia nyctalea, L.
   Common in shady, moist ground in the Minnesota valley; less common in the Sioux, James and Missouri valleys and in the Black Hills.

BORAGINACEÆ.—Borage Family.

Heliotropium Curassavicum, L. Seaside Heliotrope.
   In saline soils in the James valley in Faulk and Beadle counties.

Lappula lappula, (L.) Karst. Burrseed. European Stickseed.
Echinospermum lappula, Lehm.
Sparingly introduced in the Sioux valley.

Lappula redowskii occidentalis, (Wats.) Rydberg. Western Stickseed.
Common from the Missouri river westward.

Lappula Virginiana, (L.) Greene. Virginian Stickseed.
Echinospermum Virginicum, L.
In dry woods in the Minnesota valley and the Black Hills; rare.

Lappula floribunda, (Lehm.) Greene. Large-flowered Stickseed.
Echinospermum floribunda, Lehm.
In the Black Hills.

Lappula Americana, (Gray.) Rydberg. Nodding Stickseed.
Echinospermum deflexum Americanum, Gray.
In the Black Hills; rare.

Allocarya scopulorum, Greene. Mountain Allocarya.
In the southern Missouri valley; Charles Mix county.

Cryptanthe Pattersonii, (Gray.) Greene. Patterson's Cryptanthe.
Krynitzkia Pattersonii, Gray.
In the Black Hills; Custer and Lead City.

Cryptanthe crassisepala, (T. & G.) Greene. Thick-sepaled Cype.
Krynitzkia crassisepala, Gray.
In the Bad Lands, east of the Black Hills.

Oreocarya glomerata, (Pursh.) Greene. Clustered Oreocarya.
Krynitzkia glomerata, Gray.
In dry soils from the Minnesota valley westward.

Mertensia paniculata, (Ait.) Don. Tall Lungwort.
In the Black Hills; Rockford.

Mertensia lanceolata, (Pursh.) DC. Lance-leaved Lungwort.
Common in thickets and on plains from the Missouri valley westward.

Mertensia Sibericus, (L.) Don. Siberian Lungwort.
Dr. Rydberg says: "A single fruiting specimen which seems to belong to this species was collected at Rockford.

Myosotis macrosperma, Engelm. Large-seeded Forget-me-not.
Near Hot Springs in the Black Hills; rare.

Myosotis sylvatica, Hoffm. Sylvan Forget-me-not.
In damp, rich places in the high parts of the Black Hills.

Lithospermum canescens, (Michx.) Lehm. Hoary Puccoon.
Common on prairies and in the edge of thickets from the James valley eastward.

Lithospermum angustifolium, Michx. Narrow-leaved Puccoon.
Common on prairies throughout the state.

Onosmodium Molle, Michx. Soft-hairy False Cromwell.
In prairies throughout the state; prefers sandy soil.

Echium vulgare, L. Viper's Bugloss.
Sparingly naturalized in the Sioux valley; Brookings.

VERBENACEÆ.—Vervain Family.

Verbena urticaefolia, L. White Vervain.
In thickets in the Minnesota and Sioux valleys, and the southern part of the Missouri valley; Charles Mix county.

Verbena hastata, L. Blue Vervain.
    In thickets and moist places from the Missouri valley eastward, and in the Black Hills.

Verbena stricta, Vent. Hoary Vervain.
    In moist or dry soil throughout the state.

Verbena bracteosa, Michx. Large-bracted Vervain.
    On prairies throughout the state; rare east of the Missouri valley; more common westward.

Verbena bipinnatifida, Nutt.
    On dry soil from the Missouri valley westward.

Lippia cuneifolia, (Tott.) Steud. Fogfruit.
    Occasional in the James and Missouri valleys.

LABIATÆ.—Mint Family.

Teucrium Canadensis, L. Wood Sage. Germander.
    In moist soil in the Sioux valley.

Teucrium occidentale, Gray. Hairy Germander.
    In thickets in the Minnesota, Sioux and James valleys.

Scutellaria lateriflora, L. Mad-dog Skull Cap.
    In low, shady places from the Missouri valley eastward.

Scutellaria parvula, Michx. Small Skull Cap.
    On low prairies in the Minnesota, Sioux, James and Missouri valleys.

Scutellaria galericulata, L. Marsh Skull Cap.
    In damp ground from the Missouri valley eastward, and in the Black Hills.

Agastache nepetoides, (L.) Kuntze. Catnip. Giant Hyssop.
Lophanthus nepetoides, Benth.
    In the Minnesota valley and the southern part of the Sioux valley, at Sioux Falls; Union county. Some of the specimens from the Minnesota valley agree with *A. schrophulariæfolia* in the pubescence of the leaves

and stem, but have the small greenish yellow corolla of *A. nepetoides*.

Agastache anethoidora, (Nutt.) Britton. Fragrant Giant Hyssop.
Lophanthus anisatus, Benth.
In thickets from the Missouri valley eastward and in the Black Hills.

Nepeta cataria, L. Catnip.
Sparingly introduced in the Minnesota and Sioux valleys.

Dracocephalum parviflorum, Nutt. American Dragonhead.
In dry soils in the Sioux valley near Dell Rapids, and in the Black Hills.

Prunella vulgaris, L. Selfheal.
In damp woods in the Black Hills.

Physostegia Virginiana, (L.) Benth. False Dragonhead.
In the Minnesota, Sioux and James valleys.

Leonurus cardiaca, L. Motherwort.
Naturalized in the Sioux valley near Sioux Falls.

Stachys palustris, L. Hedge Nettle.
In the Minnesota and Sioux valleys, and in the Black Hills.

Stachys aspera, Michx. Rough Hedge Nettle.
Near Custer in the Black Hills.

Salvia lanceolata, Willd. Lance-leaved Sage.
On the dry plains from the Missouri valley westward.

Monarda fistulosa, L. Wild Bergamont.
Common in thickets from the Missouri valley eastward, and in the Black Hills.

Monarda scabra, Beck. Pale Wild Bergamont.
Monarda fistulosa Mollis, Benth.
On prairies and plains in the Minnesota valley and westward to the Black Hills.

Hedeoma hispida, Pursh. Rough Pennyroyal.
Common on dry prairies throughout the state.

Hedeoma Drummondii, Benth. Drummond's Pennyroyal.
In dry soils in the Black Hills, and in the adjacent plains.

Hyssopus officinalis, L. Hyssop.
Sparingly naturalized in the Sioux valley near Brookings.

Lycopus rubellus, Moench. Stalked Water Hoarhound.
In wet ground in the Minnesota and Sioux valleys.

Lycopus Americanus, Muhl. Cut-leaved Water Hoarhound.
Lycopus sinuatus, Ell.
Common in wet soils throughout the state.

Lycopus lucidus, Turcz. Western Water Hoarhound.
In swamps and bogs in the Minnesota and Sioux valleys.

Mentha Canadensis, L. American Wild Mint.
Common in moist soils throughout the state.

SOLANACEÆ.—Potato Family.

Physalis longifolia, Nutt. Long-leaved Ground Cherry.
In the Sioux and Missouri valleys and the Black Hills; more common westward.

Physalis lanceolata, Michx. Prairie Ground Cherry.
On dry prairies throughout the state.

Physalis Virginiana, Mill. Virginian Ground Cherry.
In thickets in the Minnesota and Sioux valleys, and in the Black Hills.

Physalis heterophylla, Nees. Clammy Ground Cherry.
In the Minnesota and southern Missouri valleys, and in the Black Hills.

Physalis rotundata, Rydb. Round-leaved Ground Cherry.
In the foothills of the Black Hills, and the adjacent plains.

Solanum nigrum, L. Black Nightshade.
In the waste places from the Missouri valley eastward and in the Black Hills.

Solanum triflorum, Nutt. Cut-leaved Nightshade.
In waste places in the Minnesota and Sioux valleys, and in the Black Hills.

Solanum rostratum, Dunal. Texas Thistle. Beaked Nightshade.
On dry prairies and in waste places throughout the state. From the Missouri valley eastward it is found in waste places, apparently introduced from the west. On the plains west of the Missouri river it is more abundant and apparently native.

SCHROPHULARIACEÆ.—Figwort Family.

Verbascum thapsus, L. Mullen.
Introduced into the southeastern part of the state, and in the Black Hills; rare.

Linaria linaria, (L.) Karst. Butter-and-eggs, Yellow Toad Flax.
Linaria vulgaris, Mill.
Sparingly naturalized in the Sioux valley.

Linaria Canadensis, (L.) Dumont. Blue Toad Flax.
In dry soils in the Black Hills.

Schrophularia Marylandica, L. Heal-all. Maryland Figwort.
Schrophularia nodosa Marylandica, Gray.
In thickets from the Missouri valley eastward, and in the Black Hills.

Pentstemon cristatus, Nutt. Crested Beard-tongue.
On dry bluffs in the Missouri valley and in the Black Hills.

Pentstemon albidus, Nutt. White Beard-tongue.
  On prairies from the Missouri valley eastward, and in the Black Hills.

Pentstemon gracilis, Nutt. Slender Beard-tongue.
  Common on prairies from the Missouri valley eastward, and in the Black Hills.

Pentstemon grandiflorus, Nutt. Large Flowered Beard tongue.
  On dry knolls and banks of ravines throughout the state.

Pentstemon glaber, Pursh. Smooth Beard-tongue.
  On low prairies from the Missouri valley westward.

Pentstemon angustifolius, Pursh. Pale Beard-tongue.
  Near Hot Springs in the Black Hills, rare.

Pentstemon Jamesii, Benth. James' Beard-tongue.
  On table-lands near Hot Springs in the Black Hills.

Collinsia parviflora, Dougl. Small-flowered Collinsia.
  On dry hill sides in the Black Hills.

Mimulus ringens, L. Monkey Flower.
  Along streams in the Minnesota and Sioux valleys.

Mimulus Jamesii, T. & G. James' Monkey Flower.
  In bogs and swamps in the Minnesota valley and in the Black Hills.

Mimulus luteus, L. Yellow Monkey Flower.
  In damp, shady ground in the Black Hills near Lead City.

Monniera rotundifolia, Michx. Round-leaved Hedge Hyssop.
  Herpestris rotundifolia, Pursh.
  In shallow pools and on muddy shores from the Missouri valley eastward, and in the Black Hills.

Ilysanthes gratioloides, (L.) Benth. False Pempernel.
  Ilysanthes riparia, Raf:
  On muddy shores from the Missouri valley eastward.

Wulfenia rubra, (Hook.) Greene. Western Wulfena.
Synthris rubra, Benth.
   On hill sides near Custer in the Black Hills.

Veronica Anagallis-aquatica, L. Water Speedwell.
   In brooks from the Missouri valley eastward and in the Black Hills.

Veronica Americana, Schwein. American Brooklime.
   In cold swamps in the Minnesota valley and in the Black Hills.

Veronica peregrina, L. Neckweed.
   In low ground from the Missouri valley eastward and in the Black Hills.

Veronica officinalis, L. Common Speedwell.
   Sparingly introduced in the Sioux valley.

Leptandra Virginica, (L.) Nutt. Culver's Root.
Veronica Virginica, L.
   On prairies in the Minnesota and Sioux valleys; rare.

Gerardia Besseyana, Britton. Bessey's Gerardia.
Gerardia tenuiflora macrophylla, Benth.
   On low prairies from the Missouri valley eastward.

Gerardia aspera, Dougl. Rough Gerardia.
   On prairies from the Missouri valley eastward; common.

Castilleja acuminata, (Pursh.) Spreng. Painted Cup.
   In woods in the Black Hills.

Castilleja sessiflora, Pursh. Prairie Painted Cup.
   On dry bluffs and sandy knolls throughout the state; much more common from the Missouri valley westward.

Orthocarpus luteus, Nutt. Yellow Orthocarpus.
   Occasional in dry, sandy soils throughout the state.

Pedicularis lanceolata, Michx. Swamp Lousewort.
   In swamps and low prairies in the Minnesota and Sioux valleys.

Pedicularis Canadensis, L. Wood Betany. Lousewort.
In thickets in the Minnesota and Sioux valleys.

LENTIBULARIACEÆ —Bladdderwort Family.

Utricularia vulgaris, L. Greater Bladderwort.
In clear water ponds and slow streams from the Missouri valley eastward.

OROBANCHACEÆ.—Broomrape Family.

Thalesia fasiculata, (Nutt.) Britton. Yellow Cancer Root.
Aphyllon fasiculatum, Gray.
In the Missouri valley and the Black Hills; rare.

Orobanche ludoviciana, Nutt. Louisiana Broomrape.
From the Missouri valley eastward and in the Black Hills.

PHRYMACEÆ.—Lopseed Family.

Phryma leptostachya, L. Lopseed.
In woops and thickets throughout the state.

PLANTAGINACEÆ.--Plantain Family.

Plantago major, L. Common Plantain.
Introduced into waste places throughout the state.

Plantago Rugelli, Dec. Rugel's Plantain.
Introduced into waste fields in the Sioux valley.

Plantago lanceolata, L. Ribwort.
Sparingly introduced into waste places in the Sioux valley.

Plantago eriopoda, Torr. Saline Plantain.
In low alkaline places in the Minnesota and Sioux valley.

Plantago Purshii, R. & S. Pursh's Plantain.
Plantago patagonica gnaphalioides, Gray.
Common on dry plains from the Missouri valley westward.

Plantago aristata, Michx. Large-bracted Plantain.
Plantago patagonica aristata, A. Gray.
On the dry plains from the Missouri valley westward; not common.

Plantago elongata, Pursh. Slender Plantain.
Plantago pusilla, Nutt.
On low damp prairies in the southern Missouri valley, Aurora and Charles Mix counties.

RUBIACEÆ.—Madder Family.

Galium Aparine, L. Cleaver. Goosegrass.
In woods and thickets in the Minnesota and Sioux valleys and in the Black Hills.

Galium boreale, L. Northern Bedstraw.
In thickets and on banks from the Missouri valley eastward and in the Black Hills.

Galium triflorum, Michx. Fragrant Bedstraw.
In woods and thickets in the Minnesota and Sioux valleys and in the Black Hills; also extends up the Missouri valley into Charles Mix county.

Galium trifidum, L. Small Cleavers.
In spring swamps in the Minnesota and Sioux valleys.

Galium tinctorum, L. Wild Madder.
In low thickets in the Sioux valley.

CAPRIFOLIACEÆ.—Honeysuckle Family.

Sambucus racemosus, L. Racemed Elder.
In canons in the Black Hills.

Sambucus canadensis, L. Sweet Elder.
In moist soil near Rapid City in the Black Hills.

Viburnum opulus, L. Cranberry Tree.
In damp ravines in the Minnesota valley (rare) and in the Black Hills.

Viburnum Lentago, L. Sweet Viburnum.
: Common on wooded banks in the Minnesota valley and in the Black Hills; occurs also near Sioux Falls in the Sioux valley.

Linnaea borealis, L. Twinflower.
: In woods in the Black Hills.

Symphoricarpus racemosus, Michx. Snowberry.
: In thickets along streams and on dry banks throughout the state.

Symphoricarpus pauciflorus, (Robbins) Britton. Low Snowberry.
: *Symphoricarpus racemosus pauciflorus.* Robbins.
: On rich wooded bluffs in the Minnesota valley and in the Black Hills.

Symphoricarpus occidentalis, Hook. Wolf Berry.
: Occasional in thickets in the Minnesota and Sioux valleys.

Symphoricarpus Symphoricarpus, (L.) MacM. Coral Berry.
: *Symphoricarpus vulgaris,* Michx.
: On dry banks from the Missouri valley eastward.

Lonicera glaucescens, Rydberg. Douglas' Honeysuckle.
: Common in woods and ravines in the Minnesota valley, in the Sioux valley near Sioux Falls and in the Black Hills.

ADOXACEÆ.—Moschatel Family.

Adoxa moschatellina, L. Moschatel. Muskroot.
: In the limestone region in the Black Hills.

VALERIANACEÆ.—Valerian Family.

Valeriana edulis, Nutt. Tobacco Root.
: In damp soil near Rockford in the Black Hills.

Valeriana sylvatica, Banks. Wood Valerian.
In moist soils in the Black Hills.

CUCURBITACEÆ.—Gourd Family.

Micrampelis lobata, (Michx.) Greene. Wild Balsam Apple.
Echinocystis lobata, T. & G.
In thickets along streams from the Missouri valley eastward.

Sicyos angulatus, L. Burr Cucumber.
In thickets along streams in the southern part of the state, east of the Missouri river; Yankton; Elk Point, and Brookings.

CAMPANULACEÆ.—Bellwort Family.

Campanula rotundifolia, L. Harebell.
On moist rocks in the Black Hills.

Campanula aparinoides, Pursh. Marsh Bellwort.
In wet ground in the Black Hills.

Campanula Americana, L. Tall Bellflower.
In thickets in the Missouri valley from Yankton southward and up the Sioux valley to Sioux Falls.

Legouzia perfoliata, (L.) Britton. Venus Looking-glass.
Specularia perfoliata, A. DC.
In the southern Missouri valley (Charles Mix county), and in the Black Hills.

Lobelia syphilitica, L. Blue Cardinal Flower.
Common in low moist ground in the Minnesota and Sioux valleys and in the Black Hills.

Lobelia spicata hirtella, Gray. Prairie Lobelia.
Common on low praires from the Missouri valley eastward.

Lobelia Kalmii, L. Brook Lobelia.
In cold spring bogs in ravines in the Minnesota valley.

CICHORIACEÆ.—Chicory Family.

Tragopogon pratensis, L.   Meadow Salsify.
Naturalized in the Minnesota and Sioux valleys.

Taraxacum Taraxacum, (L.) Karst.   Dandelion.
In lawns and and waste fields from the James valley eastward and in the Black Hills.

Sonchus asper, (L.) All.   Spiny Sow Thistle.
In waste places from the Minnesota valley eastward and in the Black Hills.

Lactuca Scariola, L.   Prickly Lettuce.
In fields and waste places from the Missouri valley eastward; not abundant.

Lactuca Ludoviciana, (Nutt.) DC.   Western Lettuce.
Common on shady banks from the Missouri valley eastward and in the Black Hills.

Lactuca Canadensis, L.   Tall Lettuce.
In most thickets in the Sioux and James valleys.

Lactuca pulchella, (Pursh.) DC.   Large-flowered Blue Lettuce.
In low places from the Missouri valley eastward and in the Black Hills.

Lactuca spicata, (Lam.) Hitch.   Tall Blue Lettuce.
Lactuca leucophæa, Gray.
In moist thickets in the Minnesota and Sioux valleys.

Lygodesmia juncea, (Pursh.) D. Don.   Rush-like Lygodesmia.
Common on breaking from the Missouri valley eastward and on dry plains from the Missouri valley westward.

Agoseris glauca, (Pursh.) Greene.   Large-flowered Agoseris.
Troximon glaucum, Pursh.
On prairies and plains throughout the state.

Agoseris parviflora, (Nutt.) Greene. Small-flowered Agoseris.
Troximon glaucum parviflorum, Gray.
In the Black Hills and the adjoining plains.

Agoseris scorsoneræfolia, (Schrad.) Greene. Western Agoseris.
On railroad embankments near Custer, Black Hills.

Nothocalai cuspidata, (Pursh.) Greene. False Calais.
Troximon cuspidatum, Pursh.
On prairies from the Missouri valley eastward.

Crepis runcinata, (James.) T. & G. Naked Stemmed Hawksbeard.
In low moist soils in the Minnesota and Sioux valleys and in the Black Hills.

Hieracium umbellatum, L. Narrow-leaved Hawkweed.
In damp ground in the Minnesota valley and in the Black Hills.

Hieracium Canadense, Michx. Canada Hawkweed.
On shady banks in the Minnesota valley and in the Black Hills.

Hieracium Fendleri, Schut. Fendler's Hawkweed.
On dry hills near Rockford in the Black Hills.

Nabalus albus, (L.) Hook. Rattlesnake Root.
Prenanthes alba, L.
In rich woods in the Minnesota valley.

Nabalus asper, (Michx.) T. & G. Rough White Lettuce.
Prenanthes asper, Michx.
On low prairies in the Minnesota, Sioux, James, and southern Missouri valleys and the Black Hills.

Nabalus racemosus, (Michx.) DC. Glaucous White Lettuce.
Prenanthes racemosa, Michx.
On low prairies in the Minnesota and Sioux valleys and in the Black Hills.

AMBROSIACEÆ.—Ragweed Family.

Iva axillaris, Pursh. Small-flowered Marsh Elder.
In dry alkaline soils from the Missouri valley westward.

Iva xanthiifolia, (Tresen.) Nutt. Burweed. Marsh Elder.
In waste places from the Missouri valley eastward and in the Black Hills.

Ambrosia trifida, L. Great Ragweed. Bitterweed.
In thickets and waste places from the Missouri valley eastward.

Ambrosia trifida integrifolia, (Muhl.) T. & G.
With the type, common.

Ambrosia artemisiaefolia, L. Ragweed.
Abundant in waste places along roads and in cultivated field from the Missouri valley eastward, rare in the Black Hills.

Ambrosia psilostachya, DC. Western Ragweed.
In light soils throughout the state; more common from the Missouri valley westward.

Gaertneria discolor, (Nutt.) Kuntze. White-leaved Gaertneria.
Franseria discolor, Nutt.
In dry soils in the Black Hills and the adjoining plains.

Xanthium Canadense, Mill. American Cocklebur.
On loose, sandy soils bordering streams and lakes from the Missouri valley eastward.

COMPOSITÆ.—Thistle Family.

Vernonia fasciculata, Michx. Western Iron Weed.
Common in low places from the Missouri valley eastward.

Eupatorium maculatum, L. Spotted Joe-Pye Weed.
In swampy ground in the Minnesota and Sioux valleys and the Black Hills.

Eupatorium maculatum amoenum, (Pursh.) Britton.
In low ground in the Minnesota and Sioux valleys. Quite distinct from the type.

Eupatorium altissimum, L. Tall Thoroughwort.
In dry soils, near timber, in the James, Sioux and Minnesota valleys and the Black Hills.

Eupatorium perfoliatum, L. Boneset.
In swamps in the Minnesota valley, in the eastern part of the Sioux valley and in the Missouri valley near Running Water.

Eupatorium ageratoides, L. F. White Snakeroot.
On wooded bluffs in the Minnesota, Sioux, and southern James valleys, and extending up the Missouri valley to Charles Mix county.

Kuhnia glutinosa, Ell. Prairie False Boneset.
Kuhnia eupatorioides corymbulosa, T. & G.
On prairies throughout the state, but rare west of the Missouri river.

Lacinaria squarrosa, (L.) Hill. Colicroot.
Liatris squarrosa, Willd.
In the southern Missouri valley; Yankton, Running Water and Charles Mix county.

Lacinaria punctata, (Hook.) Kuntze. Dotted Button Snakeroot.
Liatris punctata, Hook.
On prairies and plains throughout the state, very common. A form with creamy white flowers is found.

Lacinaria scariosa, (L.) Hill. Large Button Snakeroot. Blazing Star.
Liatris scariosa, Willd.
From the Missouri valley eastward and in the Black Hills; common.

Lacinaria spicata, (L.) Kuntze. Dense Button Snakeroot. Devil's Bit.

Liatris spicata, L.
In low prairies in the Minnesota and Sioux valleys.

Gutierrezia Sarothræ, (Pursh.) Britt. & Rusby.   Gutierrezia.
Gutierrezia Euthamiæ, T. & G.
On the dry plains from the Missouri valley westward.

Grindelia squarrosa, (Pursh.) Dunal.   Broad-leaved Gum Plant.
In dry soil throughout the state. It is becoming a weed in waste places in the eastern part of the state.

Chrysopsis villosa, (Pursh.) Nutt.   Hairy Golden Aster.
Common in dry soil from the Missouri valley eastward and in the Black Hills. Many of the western forms are very close to *C. hispida*.

Chrysothamnus Douglasii, (Gray.)   Douglas's Rayless Goldenrod.
Bigelovia Douglasii, Gray.
In dry alkaline soils from the Missouri valley westward to the Black Hills.

Eriocarpum grindelioides, Nutt.   Rayless Eriocarpum.
In dry soils in the Black Hills and the surrounding plains.

Eriocarpum spinulosum, (Nutt.) Greene.   Cut-leaved Eriocarpum.
Aplopappus spinulosus.
In dry soils throughout the state; but rare in the Minnesota and Sioux valleys, common westward.

Solidago flexicaulis, L.   Broad-leaved Goldenrod.
Solidago latifolia, L.
On shady banks in the Minnesota valley.

Solidago erecta, Pursh.   Slender Goldenrod.
On dry knolls in the Black Hills.

Solidago rigidiuscula, (T. & G.) Porter.   Slender Showy Goldenrod.
Solidago speciosa rigidiuscula, T. & G.
In dry soils in the Missouri and Sioux valleys.   Rare.

Solidago arguta, Ait.   Cut-leaved Goldenrod.
In rich woods in the Minnesota and Sioux valleys.

Solidago rupestris, Raf.   Rock Goldenrod.
Occasional on rocky banks in the Minnesota and Sioux valleys and in the Black Hills.

Solidago serotina, Ait.   Late Goldenrod.
In thickets and low places throughout the state except the Black Hills.

Solidago Missouriensis, Nutt.   Missouri Goldenrod.
On prairies throughout the state.

Solidago Canadensis, L.   Canada Goldenrod.
From the Missouri valley eastward and in the Black Hills; common in thickets and on low prairies.

Solidago Canadensis procera, (Ait.) T. & G.
Range the same as the type.

Solidago Canadensis gilvos canescens, Rydberg.
In dry soil in the Minnesota valley.

Solidago nemoralis, Ait.   Field Goldenrod.   Dyers Weed.
On dry banks and hills throughout the state.

Solidago Radula, Nutt.   Western Rough Goldenrod.
In dry soils throughout the state.

Solidago rigida, L.   Hard-leaved Goldenrod.
On prairies throughout the state.   The most abundant of all the species.

Euthamia graminifolia, (L.) Nutt.   Fragrant Goldenrod.
Solidago lanceolata, L.
In moist soils in the Minnesota, Sioux, and James valleys and in the Black Hills.

Euthamia Caroliniana, (L.) Greene. Slender Fragrant Goldenrod.
Solidago tenuifolia, Pursh.
A single specimen of the species was collected in the Black Hills by Mr. Carter in 1897.

Boltonia asteroides, (L.) L'Her. Aster-like Boltonia.
In low moist soils from the Missouri valley eastward.

Aster Lindleyanus, T. & G. Lindley's Aster.
In low ground in the Minnesota valley.

Aster sagittifolius, Willd. Arrow-leaved Aster.
In dry soils in the Minnesota valley.

Aster Novæ-Angliæ, L. New England Aster.
In thickets in the Minnesota, Sioux, and James valleys.

Aster oblongifolius, Nutt. Aromatic Aster.
In dry soils from the Missouri valley eastward; the plant is most abundant on dry gravelly hillsides.

Aster lævis, L. Smooth Aster.
On sandy banks of streams from the Missouri valley eastward and in the Black Hills.

Aster patulus, Lam. Spreading Aster.
In the Black Hills near Custer.

Aster junceus, Ait. Rush Aster.
In cold spring bogs in the Minnesota valley, in the extreme eastern part of the Sioux valley and in the Black Hills.

Aster sericeus, Vent. Western Silky Aster.
On dry prairies in the Minnesota, Sioux, and James valleys; common.

Aster ptarmicoides, (Nees.) T. & G. Upland White Aster.
In dry rocky or gravelly soils throughout the state.

Aster dumosus, L. Bushy Aster.
In damp, sandy soils in the Sioux valley.

Aster salicifolius, Lam. Willow Aster.
In low wet ground in the vicinity of streams, throughout the state.

Aster paniculatus, Lam. Panicled Aster.
In low, damp ground from the Missouri valley eastward.

Aster lateriflorus, (L.) Britton. Starved Aster.
Aster diffussis, Ait.
On shady banks in the Sioux valley.

Aster multiflorus, Ait. Dense-flowered Aster.
In dry soils, especially that have been broken, from the Missouri valley eastward.

Aster incanopilosus, (Lindl.) Sheldon. White Prairie Aster.
Aster commutatus, A. Gray.
On dry prairies throughout the state.

Aster Sibiricus, L. Siberian Aster.
Near Custer in the Black Hills.

Aster falcatus, Lindl.
Reported for the Black Hills in Gray's list, Newton and Jenney's report in the Geological survey of the Black Hills in 1880.

Machæranthera sessiliflora, (Nutt.) Greene. Viscid Aster.
On the dry plains from the Missouri valley westward.

Erigeron asper, Nutt. Rough Erigeron.
Erigeron glabellus, Nutt.
In dry soils in the Black Hills.

Erigeron subtrinervis, Rydberg. Three-nerved Fleabane.
On shaded hillsides in the Black Hills.

Erigeron pumilus, Nutt. Low Erigeron.
On the dry plains from the Missouri valley westward.

Erigeron canus, Gray.   Hoary Erigeron.
   In dry soils in the Black Hills and the adjacent plains.

Erigeron compositus, Pursh.   Dwarf Fleabane.
   On exposed soils in the Black Hills.

Erigeron flagellaris, Gray.   Running Fleabane.
   In rich soils in the Black Hills.

Erigeron salsuginous, Gray.
   In dry soils in the foothills of the Black Hills and the adjacent plains.

Erigeron pulchellus, Michx.   Robin's Plantain.
   Erigeron bellidifolius, Muhl.
   On banks of streams in the Minnesota and Sioux valleys.

Erigeron Philadelphicus, L.   Skevish.
   In low ground in the Minnesota, Sioux, and James valleys and in the Black Hills.

Erigeron annuus, (L.) Pers.   Sweet Scabious.
   On low prairies in the Minnesota and Sioux valleys.

Erigeron ramosus, (Walt.) B. S. P.   Daisy Fleabane.
   Erigeron strigosus, Muhl.
   In low ground and on banks throughout the state.

Erigeron ramosus Beyrichii, (F. & M.) Smith & Pound.
   In the Black Hills.

Erigeron armerifolius, Turcz.   Mountain Fleabane.
   In wet meadows in the Black Hills.

Leptilon Canadense, (L.) Britton.   Horseweed.
   Erigeron Canadensis, L.
   Throughout the state; common in the eastern part of the state; rare from the Missouri valley westward.

Leptilon divaricatum, (Michx.) Raf.   Purple Horseweed.
   Erigeron divaricatus, Michx.
   In sandy soils in the Missouri valley in Walworth county, rare.

Doellingeria umbellata, (Mill.) Nees.   Flat-topped White Aster.
Aster umbellatus, Mill.
In swampy places in shaded ravines in the Minnesota valley.

Filago prolifera, (Nutt.)   Britton.   Filago.
Evax prolifera, Nutt.
On dry hills near Hot Springs in the Black Hills.

Antennaria dioica, (L.) Gaertn.   Mountain Everlasting.
In the borders of open woods in the Black Hills.

Antennaria neodioica, Greene.   Smaller Cats-foot.
In shady places in the Black Hills. Two of Dr. Rydberg's specimens; 795 collected near Hermosa, 793 near Hot Springs are also referred to this species. They differ from the type in having the leaves tomentose above.

Antennaria campestris, Rydberg.   Prairie Cats-foot.
On prairies throughout the state; this is the common species which has been regarded as a form of *A. plantaginifolia* which has not yet been collected in the state. It is likely that it occurs in the open woods in the Minnesota and Sioux valleys.

Antennaria Aprica, Greene.
In the Bad Lands.

Antennaria parvifolia, Nutt.
Along Spring Lake in Brown county.

Anaphalis margaritacea, (L.) .B. & H.
Antennaria margaritacea, Hook.
In dry soils in the Black Hills.

Silphium perfoliatum, L.   Cup Plant.
In moist thickets in the Minnesota and Sioux valleys.

Silphium laciniatum, L.   Compass Plant.
In the southern part of the Sioux, James and Missouri valleys. Sioux Falls, Yankton, Charles Mix counties.

Heliopsis scabra, Dunal. Rough Ox Eye.
Common in thickets and open woods from the Missouri valley eastward. Occasional westward along the White and Bad rivers.

Rudbeckia hirta, L. Black-eyed Susan.
On prairies in the Minnesota and Sioux valleys and in the Black Hills.

Rudbeckia laciniata, L. Green-headed Coneflower.
In thickets in the Minnesota, Sioux, and southern Missouri valleys as far north as Charles Mix county.

Ratibida pinnata, (Vent.) Barnhart. Gray-headed Coneflower.
Lepachys pinnata, T. & G.
On low prairies in the Sioux valley.

Ratibida columnaris, (Sims.) D. Don. Prairie Coneflower.
Lepachys columnaris, T. & G.
On prairies from the Missouri valley eastward and in the Black Hills.

Brauneria pallida, (Nutt.) Britton. Pale Purple Coneflower.
Echinacea angustifolia, DC.
Common on prairies from the Missouri valley eastward and in the Black Hills.

Helianthus annuus, L. Common Sunflower.
Common throughout the state.

Helianthus petiolaris, Nutt. Prairie Sunflower.
On dry prairies throughout the state; rare from the James valley eastward.

Helianthus scaberrimus, Ell. Stiff Sunflower.
Helianthus rigidus, Desf.
Abundant on dry prairies throughout the state.

Helianthus Maximiliani, Schrad. Maximilian's Sunflower.
On rather low prairies throughout the state.

Helianthus grosse-serratus, Marteus. Saw-toothed Sunflower.
  In low ground in the Minnesota and Sioux valleys; common.

Helianthus tuberosus, L. Jerusalem Artichoke.
  In moist soils in the vicinity of streams, from the Missouri valley eastward.

Helianthus tuberosus subcanescens, Gray.
  In the Sioux valley; rare.

Helianthella quinquenervis, (Hook.) Gray. Five-nerved False Sunflower.
  On dry knolls in the Black Hills.

Balsamorhiza sagittata, (Pursh.) Nutt. Balsamroot.
  In the Black Hills.

Coreopsis tinctoria, Nutt. Garden Tickseed.
  In moist soils from the Missouri valley eastward.

Cereopsis palmata, Nutt. Stiff Tickseed.
  In the Sioux valley near Sioux Falls; rare.

Bidens laevis, (L.) B. S. P. Smooth Burr Marigold.
  In wet meadows in the Black Hills.

Bidens cernua, L. Nodding Burr Marigold.
  In wet soils from the Missouri valley eastward.

Bidens connata, Muhl. Purple-stemmed Beggartick.
  In swamps in the Sioux valley.

Bidens frondosa, L. Sticktight.
  In moist soils from the Missouri valley eastward.

Hymenopappus tenuifolius, Pursh. Woolly Hymenopappus.
  On dry plains from the Missouri valley westward.

Hymenopappus filifolius, Hook. Tufted Hymenopappus.
  On dry prairies from the Missouri valley westward.

Bahia oppositifolia, Nutt. Bahia.
  On dry plains from the Missouri valley westward.

Picradenia acaulis, (Nutt.) Britton.   Stemless Picradenia.
Actinella acaulis, Nutt.
   In dry, gravelly or sandy soils in the Black Hills, and on the hills and buttes from the Missouri valley westward.

Helenium autumnale, L.   Sneezewort.
   In low, wet places in the Minnesota, Sioux and James valleys.

Gaillardia aristata, Pursh.   Great Flowered Gaillardia.
   On dry gravelly banks and knolls in the Minnesota valley and the Black Hills.

Dysodia papposa, (Vent.) A. S. Hitchcock.   Fetid Marigold.
Dysodia chrysanthemoides, Lag.
   In low places along streams from the James valley westward. In many places in the Missouri valley it has become a bad weed.

Achillea Millefolium, L.   Yarrow.
   In fields and waste places in the Minnesota and Sioux valleys, and in the Black Hills.

Anthemis Cotula, L.   Mayweed.
   In waste places from the Missouri valley eastward and in the Black Hills.

Chrysanthemum Leucanthemum, L.   White Daisy.
   Sparingly introduced in the Sioux valley.

Tanacetum vulgare, L.   Tansy.
   Escaped from gardens in the Sioux and James valleys, rare.

Artemisia Canadensis, Michx.   Canada Wormwood.
   On sandy hillsides in the Black Hills.

Artemisia caudata, Michx.   Wild Wormwood.
   In sandy soils throughout the state.

Artemisia dracunculoides, Pursh.   Common Wormwood.
   In thickets and on prairies throughout the state.

Artemisia filifolia, Torr.   Silvery Wormwood.
In the Bad Lands country just east of the Black Hills.

Artemisia frigida, Willd.   Wormwood Sage.
In loose gravelly soils throughout the state.

Artemisia biennis, Wild.   Biennial Wormwood.
Introduced from the Missouri valley eastward. A bad weed, native of the Northwest Territory.

Artemisia serrata, Nutt.   Saw-leaved Mugwort.
In thickets and on low ground, not common.

Artemisia longifolia, Nutt.   Long-leaved Mugwort.
On the dry plains from the Missouri valley westward, occasional in the James valley.

Artemisia gnaphalodes, Nutt.   Prairie Mugwort.
Abundant on prairies throughout the state.

Artemisia cana, Pursh.   Hoary Sagebrush.
On the dry plains from the Missouri valley westward.

Petasites sagittata, (Pursh.) Gray.   Sweet Coltsfoot.
In the Black Hills near Rochford.

Arnica cordifolius, Hook.   Heart-leaved Arnica.
On shady hillsides in the Black Hills.

Arnica alpina, (L.) Olin.   Arctic Arnica.   Mountain Tobacco.
In canons and on shady banks in the Black Hills.

Senecio integerrimus, Nutt.   Entire-leaved Groundsel.
Occasional in the Minnesota and Sioux valleys and in the Black Hills.

Senecio lugens, Richards.   Black Tipped Groundsel.
On prairies in the Sioux and James valleys and in the Black Hills.

Senecio canus, Hook.   Silvery Groundsel.
In dry soils in the Black Hills, a common and variable species.

Senecio Plattensis, Nutt. Prairie Ragwort.
In dry ground in the Black Hills.

Senecio Balsamitæ, Muhl. Balsam Groundsel.
Senecio aureus Balsamitæ, T. & G.
On prairies from the Missouri valley eastward and in the Black Hills; common.

Senecio aureus, L. Golden Ragwort.
In low, wet ground in the Minnesota and Sioux valleys.

Senecio Douglasii, DC. Douglas' Ragwort.
In dry soils in the plains region, along the Cheyenne river.

Senecio palustris, (L.) Hook. Marsh Ragwort.
In swamps in the Minnesota and Sioux valleys.

Senecio vulgaris, L. Common Groundsel.
Occurs in cultivated and waste ground in the Minnesota and Sioux valleys.

Senecio eremophilus, Richards. Mountain Ragwort.
On dry soils in the Black Hills.

Senecio rapifolius, Nutt.
In shady places in the Black Hills.

Senecio discoideus, (Hook.) Britton. Northern Squaw Weed.
In moist ground in the Missouri valley, Charles Mix county. The specimens are typical, except that the achenes are sharply four-sided and hairy on the angles.

Arctium minus, Schk. Common Burdock.
Sparingly naturalized in shady waste places in the southern part of the Sioux valley and in the Minnesota valley.

Carduus altissimus, L. Roadside Thistle.
Cnicus altissimus, Willd.
In fields, woods and waste places in the Minnesota and Sioux valleys.

Carduus discolor, (Muhl.) Nutt. Field Thistle.
Cnicus discolor, Muhl.
On prairies in the Minnesota and Sioux valleys.

Carduus Virginianus, L. Virginian Thistle.
Cnicus Virginianus, Pursh.
In woods and thickets in the Minnesota and Sioux valleys.

Carduus undulatus, Nutt. Wavy-leaved Thistle.
Cnicus undulatus, Gray.
On plains and prairies throughout the state.

Carduus ochrocentrus, (Gray.) Green. Yellow-spined Thistle.
Cnicus ochrocentrus, Gray.
In dry soils in the Black Hills.

Carduus Plattensis, Rydberg. Prairie Thistle.
In sandy soils in the southwestern part of the state.

Carduus Drummondi, (Gray).
Cnicus Drummondi, Gray.
In damp meadows in the Black Hills.

Carduus Carnovirens, Rydberg.
Collected near Sylvan lake in the Black Hills, by Mr. L. W. Carter. Specimens sent Dr. Rydberg were identified as this species.

Carduus arvensis, (L.) Robs. Canada Thistle.
Cnicus arvensis, Hoff.
Sparingly naturalized in the Sioux valley.

Centaurea Cyanus, L. Blue Bottle. Corn Flower.
Roadsides near Hot Springs in the Black Hills.

INDEX.

	PAGE
Abronia	141
Abutilon	171
ACERACEÆ	169
Acer	169
Acerates	183
Achillea	210
Acnida	140
Aconitum	144
Acorus	125
Actinella	210
Actæa	144
Acuan	158
Adder-Tongues	103
Adiantum	104
Adicea	135
ADOXACEÆ	196
Adoxa	196
Agastache	188
Agoseris	198
Agrimonia	156
Agropyron	118
Agrostemma	142
Agrostis	113
Alexander	178
Alisma	108
ALISMACEÆ	108
Allium	127
Allionia	140
Allocarya	186
Alopecurus	112
Alsine	141
Alum-root	152
AMARANTHACEÆ	140
Amaranths	140
Amaranthus	140
AMARYLLIDACEÆ	130
Amaryllis	130
AMBROSIACEÆ	200
Ambrosia	200
Amelanchier	157

	PAGE
American Brooklime	193
American Mint	190
American Spikenard	176
Ammannia	174
Amorpha	160
Ampelopsis	171
Amphicarpa	165
ANACARDIACEÆ	169
Anaphalis	207
Andropogon	109
Androsace	181
Anemone	144
ANGIOSPERMS	106
Anogra	175
Antennaria	207
Anthemis	210
Aphyllon	194
Apios	165
Aplopappus	202
APOCYNACEÆ	183
Apocynum	183
Apple Family	156
Aquilegia	144
Arabis	151
ARACEÆ	125
ARALIACEÆ	176
Aralia	176
Arctium	212
Arctostapylos	180
Arenaria	143
Argemone	147
Arisæma	125
Aristida	111
Arnica	211
Arrhenatherum	114
Arrow-grass	108
Arrowhead	108
Artemisia	210
Artichoke	209
Arums	125

	PAGE		PAGE
ASCLEPIADACEÆ	183	Black-eyed Susan	208
Asclepias	183	Black Mustard	148
Ash	181	Black Raspberry	154
Asparagus	128	Black Snakeroot	177
Aspen	132	Black Walnut	131
Asplenium	104	Bladder Campion	142
Aster	204	Bladder-pod	150
Astragalus	161	Bladderwort	194
Atriplex	139	Blazing Star	201
Avena	114	Blight	139
Avens	156	Blitum	139
Bahia	209	Bloodroot	147
BALSAMINACEÆ	170	Blue Cohosh	147
Balsamorrhiza	209	Blue-eyed Grass	130
Balsam Poplar	132	Blue Joint	114
Balsamroot	209	Boltonia	204
Baneberry	144	Boneset	201
Barberry	147	BORAGINAGEÆ	185
Barnyard-grass	109	Borage Family	185
Basswood	171	Botrychium	103
Bastard Toad-flax	135	Bottle Brush	176
Batrachium	146	Bouncing Bet	141
Bearberry	180	Bouteloua	115
Beard-grass	109	Box Elder	170
Beard-tongues	192	Brachyelytrum	112
Bear's-grass	128	Brake	104
Beckmannia	115	Brassica	148
Beckwith's Clover	159	Brauneria	208
Bedstraws	195	Brewer's Cliff-brake	104
Bellwort Family	197	Brittle-fern	103
Bellwort	127	Brome	118
Bent-grass	113	Bromus	118
BERBERIDACEÆ	147	Broomrape Family	194
Berberis	147	Buckbean Family	182
Berula	179	Buckthorn Family	170
BETULACEÆ	133	Buckwheats	135
Betula	133	Buffalo Berry	174
Bicuculla	147	Buffalo Burr	164
Bidens	209	Buffalo Clover	159
Bigelovia	202	Buffalo Currant	153
Bilberry	180	Buffalo Grass	115
Bindweed	184	Buffalo Pea	161
Birch	133	Bug-seed	139
Bird's-foot Trefoil	160	Bulbilis	115
Bittersweet	169	Bulrush	121
Bitterweed	200	Bunch-flowers	127
Black Cherry	158	Bupleurum	179

	PAGE		PAGE
Burdock	212	Catnip	188
Bur-head	108	Cat's-foot	207
Burning Bush	169	Cat-tails	106
Burr Cucumber	197	Caulophyllum	147
Bur-reed	106	Ceanothus	170
Burr Marigold	209	CELASTRACEÆ	169
Burr Oak	134	Celastrus	169
Bursa	150	Celtis	134
Bush Clover	164	Centaurea	213
Bushy Blue-stem	109	Centunculus	181
Buttercups	144	Cerastium	142
Butterfly Weed	183	CERATOPHYLLACEÆ	143
Button Snakeroot	201	Ceratophyllum	143
CÆSALPINACEÆ	158	Cercis	158
CACTACEÆ	173	Cercocarpus	156
Cactus Family	173	Chaffweed	181
Cactus	173	Chamænerion	174
Calamagrostis	114	Cheeses	171
Calamovilfa	114	Cheilanthes	104
CALLITRICHACEÆ	168	Cenchrus	110
Callitriche	168	CHENOPODIACEÆ	138
Calochortus	128	Chenopodium	138
Caltha	144	Cherries	157
Camelina	150	Chickweed	141
CAMPANULACEÆ	197	Chicory Family	198
Campanula	197	Choke Cherry	158
Canary Grass	110	Chrysanthemum	210
Cancer Root	194	Chrysopogon	109
Canoe Birch	133	Chrysothamnus	202
Cannabis	134	CICHORIACEÆ	198
Caper Family	152	Cicuta	179
Capnoides	148	Cinna	113
CAPPARIDACEÆ	152	Cinquefoils	154
CAPRIFOLIACEÆ	195	Circæa	176
Capsella	150	CISTACEÆ	172
Caraway	179	Clammy-weed	152
Cardamine	149	Claytonia	141
Carduus	212	Clearweed	135
Carex	122	Cleavers	195
Carrion Flower	129	Clematis	145
Carrot Family	177	Cleoma	152
Carum	179	Cliff-brake	104
CARYOPHYLLACEÆ	141	Clovers	159
Cassia	158	Club Mosses	105
Castilleja	193	Club Rushes	121
Catabrosa	116	*Cnicus*	213
Catchfly	141	Cocklebur	200

	PAGE		PAGE
Colicroot	201	CUSCUTACEÆ	184
Collinsia	192	Cuscuta	184
Collomia	185	Cup Plant	207
Columbine	144	Currant	153
Comandra	135	Cycloloma	139
COMMELINACEÆ	126	Cymopterus	177
Compass Plant	207	CYPERACEÆ	120
COMPOSITÆ	200	Cyperus	120
CONVALLARIACEÆ	128	Cypripedium	130
CONVOLVULACEÆ	184	Cystopteris	103
Convolvulous	184	Dactylis	116
Corallorhiza	131	Dakota Vetch	160
Coral-root	131	*Dalea*	161
Cord-grass	115	Dandelion	198
Coreopsis	209	Danthonia	115
Coringia	152	Daucus	177
Corispernum	139	Delphinium	144
CORNACEÆ	179	Deringa	179
Corn Cockle	142	*Desmanthus*	158
Cornel	179	*Desmodium*	164
Cornflower	213	Dicentra	147
Cornus	179	DICOTYLEDONES	131
Corydalis	148	Diplachne	115
Corylus	133	Diosporum	129
Cottonwood	132	Distchlis	116
Cotton-grass	121	Docks	136
Couch-grass	118	Dodder Family	184
Cow-herd	141	Dodecatheon	181
Cow-parsnip	177	Doellingeria	207
Cowslip	181	Dogbane Family	183
Crab-grass	109	Dogwood Family	179
Cranberry Tree	195	Dogwoods	179
Crane's Bills	166	Dondia	139
CRASSULACEÆ	152	Draba	150
Cratægus	157	Dracocephalum	189
Creeping Red Cedar	106	Dragonhead	189
Crepis	199	DRUPACEÆ	157
Cress	149	Dryopteris	104
Crotalaria	159	Duckweeds	126
Croton	167	Dutchman's Breeches	147
Crowfoots	144	Dwarf Cornel	179
CRUCIFERÆ	148	Dysodia	210
Cryptotænia	179	Eatonia	116
Cryptanthe	186	*Echinacea*	208
Cuckoo-flower	150	*Echinocystis*	197
CUCURBITACEÆ	197	Echinodorus	108
Culver's Root	193	*Echinospermum*	186

	PAGE
Echium	187
ELATINACEÆ	172
Elatine	172
Elder	195
ELAEAGNACEÆ	174
Elaeagnus	174
Eleocharis	120
Eel-grass	108
Ellisia	185
Elm	134
Elymus	119
Enchanter's Nightshade	176
Epilobium	174
Equisetum	105
EQUISETACEÆ	105
Eragrostis	116
ERICACEÆ	180
Erigeron	205
Eriocarpum	202
Eriogonum	135
Eriophorum	121
Eryngium	177
Erysimum	151
Euonymus	169
Eupatorium	200
EUPHORBIACEÆ	167
Euphorbia	167
Eurotia	139
Euthamia	203
Evax	207
Evening Primrose Family	174
Evening Primrose	174
Evolvulus	184
FAGACEÆ	134
Falcata	165
False Indigo	160
False Pimpernel	192
False Red-top	117
Fame-flower	141
Ferns	103
Fescue	118
Festuca	118
Fetid Marigold	210
Figwort Family	191
Filago	207
Finger Grass	109
Fireweed	174

	PAGE
Flag	130
Flax Family	166
Fleabanes	205
Flowering Plants	106
Fogfruit	187
Forget-me-not	187
Four-o'clocks	140
Foxtail	110
Fragaria	154
Franseria	200
Frasera	182
Fraxinus	181
Fringed Orchis	131
Fritillaria	128
Frostweed	172
Gaertneria	200
Gaillardia	210
Galium	195
Garlics	128
Gaura	176
Gayophyton	176
GENTIANACEÆ	182
Gentiana	182
Gentian Family	182
Gentians	182
GERANIACEÆ	166
Geranium Family	166
Geranium	166
Gerardia	193
Germander	188
Geum	156
Giant Hyssop	188
Gilia	185
Ginseng Family	176
Glasswort	139
Gleditsia	158
Glyceria	117
Glycyrrhiza	164
Golden Aster	202
Goldenrods	203
Goodyera	131
Gooseberry	153
Goosefoots	138
Goosegrass	194
Gourd Family	197
GRAMINEÆ	109
Grape Family	170

Grape-fern	103
Grapes	170
Grasses	109
Grass of Parnassus	152
Green Ash	181
Greenbrier	130
Grindelia	202
Gromwell	187
GROSSULARIACEÆ	153
Ground Cherry	192
Groundnut	165
Ground Pine	105
Ground Plum	161
Groundsel	211
Gum Plant	202
Gutierrezia	202
Gymnocladus	158
GYMNOSPERMÆ	105
Gymnosperms	105
Gyrostachys	131
Habenaria	130
Hackberry	134
HALORAGIDACEÆ	176
Harebell	197
Hare's-ear Mustard	152
Hazelnut	133
Hawkweed	199
Heath Family	180
Hedeoma	189
Hedge Bindweed	184
Hedge Hyssop	192
Hedge Mustard	148
Hedge Nettle	189
Hedysarum	164
Helenium	210
Helianthella	209
Helianthemum	172
Helianthus	208
Heliopsis	208
Heliotropium	185
Hemp	134
Heracleum	177
Herpestis	192
Heteranthera	126
Heuchera	152
Hibiscus	171
Hieracium	199
Hierochloe	111
Hippuris	176
Hoarhound	190
Hog Peanut	165
Homalobus	163
Homalocenchrus	110
Honewort	179
Honeysuckle Family	195
Hop Clover	159
Hops	134
Hordeum	119
Hornwort	143
Hosackia	162
Horseradish	149
Horse-tails	105
Horseweed	206
Huckleberry Family	180
Humulus	134
Hungarian Grass	110
HYDROPHYLLACEÆ	185
Hydrophyllum	185
Hymenopappus	209
HYPERICACEÆ	171
Hypericum	171
Hypoxis	130
Hyssop	189
Hyssopus	190
Hysanthes	192
Impatiens	170
Indian Hemp	183
Indian Mallow	171
Indian Pipe Family	180
Indian Pipe	180
Indian Rice	110
Indian Soapweed	128
Indian Turnip	125
Ipomœa	184
IRIDACEÆ	130
Iris	130
Ironweed	200
Iron-wood	133
Iva	200
Ixophorus	110
Jewel-weed Family	170
JUNCACEÆ	126
Juncus	126
Juncoides	127

	PAGE		PAGE
JUGDLANDACEÆ	131	LINACEÆ	166
Juglans	131	Linaria	191
June Berry	157	Linden Family	171
Juniper	106	Linnaea	196
Juniperus	106	Linum	166
Kentucky Blue-grass	116	Lip-fern	104
Kentucky Coffee-tree	158	Lippia	187
Knot-grass	137	Little Club Mosses	105
Knot-weed	137	LOASACEÆ	173
Koeleria	116	Loasa Family	173
Krynitzkia	186	Lobelia	197
Kuhnia	201	Loco-weed	163
Kunistera	161	Lonicera	196
LABIATÆ	188	Loosestrife Family	174
Laciniaria	201	Loosestrife	174
Lactuca	198	Lophanthus	189
Ladies' Slipper	130	Lophotocarpus	108
Ladies' Tresses	131	Lopseed	194
Lady-fern	104	Lotus	160
Lamb's quarters	138	Lousewort	193
Laportea	135	Luetkea	154
Lappula	186	Lungwort	187
Larkspur	144	Lupines	159
Lathyrus	165	Lupinus	159
Lead Plant	160	Lychnis	141
Leeks	127	LYCOPODIACEÆ	105
Leersia	110	Lycopus	190
Legouzia	197	Lycopodium	105
Lemna	126	Lygodesmia	198
LEMNACEÆ	126	*Lysimachia*	181
LENTIBULARIACEÆ	194	LYTHRACEÆ	174
Leonurus	189	Lythrum	174
Lepachys	208	Machæranthera	205
Lepargyræa	174	Macrocalyx	185
Lepidium	148	Madder Family	194
Leptandra	193	Mahonia	147
Lepti'on	206	Maiden-hair	104
Lespedeza	164	*Majanthemum*	129
Lesquerella	150	Male-fern	104
Lettuce	198	Mallow Family	171
Leucocrinum	527	Malva	171
Liatris	201	MALVACEÆ	171
Lithospermum	187	Malvastrum	171
LILIACEÆ	127	*Mamillaria*	173
Lilies	127	Manna Grass	117
Lilium	128	Maple Family	169
Lily of the Valley	128	Mariposa Lily	128

	PAGE		PAGE
Marsh Foxtail	112	Moss Champion	142
Marsh Marigold	144	Moschatel Family	196
Marsilea	104	Motherwort	189
MARSILEACEÆ	104	Mountain Ash	157
Matthiola	151	Mountain Rice	111
Matricary Grape fern	103	Mouse Tail	145
Mayweed	210	Mud Plantain	126
Meadow Grass	117	Muhlenbergia	112
Meadow Parsnip	178	Mulberry	134
Meadow Rue	147	Mullen	192
Meadow-sweet	154	Munroa	115
Meibomia	164	Musineon	178
MELANTHACEÆ	127	Mustard Family	148
Melilotus	159	Myosotis	187
MENISPERMACEÆ	147	Myosurus	145
Menispermum	147	Myriophyllum	176
Mentha	190	Nabalus	199
Mentzelia	173	Naiad	106
MENYANTHACEÆ	182	NAIADACEÆ	106
Menyanthes	182	Naias	107
Meriolix	175	*Nasturtium*	149
Mertensia	187	Naumbergia	181
Mesquite Grass	115	Neckweed	193
Micrampelis	197	Nepeta	189
Milk Purslane	167	Nettles	135
Milk Vetches	162	Nightshades	191
Milkweed Family	183	Ninebark	153
Milkwort Family	167	Northern spleanwort	104
MIMOSACEÆ	158	Nothocalia	198
Mimosa Family	158	NYCTAGINACEÆ	140
Mimulus	192	Nymphæa	143
Mint Family	188	NYMPHÆACEÆ	143
Modesty	179	Oak-fern	103
Mœhringia	143	Oat-grass	114
Monarda	189	Œnothera	175
MONOCOTYLEDONS	106	OLEACEÆ	181
Monkey Flower	192	Oleaster Family	174
Monkshood	144	Olive Family	181
Monniera	192	Onagra	175
Monolepis	139	ONAGRACEÆ	174
Monotropa	180	Onoclea	103
MONOTROPACEÆ	180	Onosmodium	187
Moonseed Family	147	OPHIOGLOSSACEÆ	103
MORACEÆ	134	Opulaster	153
Morning Glory Family	184	Opuntia	173
Morongia	158	Orchard-grass	116
Morus	134	ORCHIDACEÆ	130

	PAGE
Orache	139
Oreocarya	186
Oregon Woodsia	103
Orchids	130
OROBANCHACEÆ	194
Orobanche	194
Orophaca	163
Orpine Family	152
Orthocarpus	193
Oryzopsis	111
Osmorrhiza	178
Ostrich-fern	103
Ostrya	133
OXALIDACEÆ	166
Oxalis	166
Ox Eye	208
Oxygraphis	146
Oxytropis	163
Pachylophus	175
Painted Cup	193
Panicularia	117
Panicum	109
PAPAVERACEÆ	147
PAPILLIONACEÆ	158
Parietaria	135
Parnassia	152
Paronichia	143
Parosela	161
Parsley	177
Parsnip	178
Parthenocissus	170
Pasque Flower	145
Pastinaca	177
Pea Family	158
Pedicularis	193
Pellaea	104
Pellitory	135
Pennyroyal	189
Penthorum	152
Pentstemon	191
Pepper-grass	148
Peramium	131
Persicaria	136
Petalostemon	161
Petasites	211
Peucedanum	177
Phaca	163

	PAGE
Phalaris	110
Phegoptoris	103
Philotria	108
Phleum	112
Phlox	185
Phlox Family	185
Phragmites	115
Phryma	194
PHRYMACEÆ	194
Physalis	190
Physaria	150
Physostegia	189
Picea	105
Pickerel-weed	126
Picradenia	210
Pigeon grass	110
Pigweed	138
Pilea	135
PINACEÆ. Pinus	105
Pine Drops	180
Pinks	141
PLANTAGINACEÆ	194
Plantago	194
Plantain Family	194
Pleurisy Root	183
Plum Family	157
Poa	116
Poison Ivy	169
Poison Oak	169
Polanisia	152
POLEMONIACEÆ	185
Polygala	167
POLYGALACEÆ	167
POLYGONACEÆ	135
Polygonatum	129
Polygomum	136
POLYPODIACEÆ	103
Polypodium	104
Polypody	104
POMACEÆ	156
Pond Lily	143
Pond Weed	106
Poplar	132
Poppy Family	147
Populus	132
Porcupine Grass	111
Portulaca	141

	PAGE		PAGE
PORTULACACEÆ	141	Reed Grass	113
Potamogeton	106	RHAMNACEÆ	170
Potato Family	190	Rhus	169
PONTEDERIACEÆ	126	Ribes	153
Potentilla	154	Rice Cut-Grass	110
Powder-horn	142	Rock-cress	151
Prairie Clover	161	Rock Maple	169
Prairie Turnip	160	Rocky Mountain Woodsia	103
Prenanthes	199	Rock-rose Family	172
Prickly Ash	167	Roripa	148
Prickly Pear	173	Rosa	156
Primrose Family	181	ROSACEÆ	153
Prosartes	129	Rose Family	153
PRIMULACEÆ	181	Roses	156
Prunella	189	Rudbeckia	208
Prunus	157	RUBIACEÆ	194
Psoralea	160	Rubus	154
PTERIDOPAYTES	103	Rue Family	167
Pteris	104	Rumex	136
Pterospora	180	Ruppia	107
Puccoon	187	Rushes	126
Purple-stemed Cliff-brake	104	Rush-grass	112
Purslane	141	Russian Thistle	139
Pursley	141	RUTACEÆ	167
Pulsatilla	145	Sage	189
Pyrola	180	Sagebush	211
PYROLACEÆ	180	Sagittaria	108
Pyrus	157	SALICACEÆ	132
Quack-grass	118	Salicornia	139
Quamoclit	184	Salix	132
Quercus	134	Salmon-berry	154
Ragweed Family	200	Salsify	198
Ragworts	212	Salsola	139
RANUNCULACEÆ	144	Salvia	189
Ranunculus	146	Sambucus	195
Ratibida	208	Sand Cherry	157
Rattle-box	159	Sand Burr	110
Rattlesnake Plantain	131	Sandwort	143
Rattlesnake Root	199	Sanguinaria	147
Rayless Goldenrod	202	Sanicula	177
Red Ash	181	SANTALACEÆ	135
Redbud	158	Saponaria	141
Red Cedar	106	Sarsaparilla	177
Red Osier	179	Savastana	111
Red Raspberry	154	Saxifraga	152
Redroot	170	SAXIFRAGACEÆ	152
Redtop	113	Saxifrages	152

	PAGE
Schedonnardus	115
SCHEUCHZERIACEÆ	108
Schollera	126
Schrankia	158
Schrophularia	191
SCHROPHULARIACEÆ	191
Scirpus	121
Scouring-rush	105
Scutellaria	188
Sedges	120
Sedum	152
Selaginella	105
SELAGINELLACEÆ	105
Selfheal	189
Senaca Grass	111
Senecio	211
Senna Family	158
Sensitive-brier	158
Sensitive fern	103
Senitive Pea	158
Service Berry	157
Setaria	110
Sheep Sorrel	136
Shepherdia	174
Shepherd's Purse	150
Shin-leaf	180
Shoe-string	161
Shooting Star	181
Sickle-pod	151
Sicyos	197
Silene	141
Silphium	207
Silver-berry	174
Silver Maple	169
Sinapsis	148
Sisymbrium	148
Sisyrinchium	130
Sium	178
Skullcap	188
Skunk-bush	169
Slippery Elm	134
Smartweed	137
SMILACACEÆ	129
Smilacina	128
Smilax	129
Snakeroot	177
Sneezewort	210

	PAGE
Snowberry	196
Soapwort	141
Soft Maple	169
SOLANACEÆ	190
Solanum	191
Solidago	202
Solomon's Seal	129
Sonchus	198
Sophia	150
Sophora	158
Sorbus	156
Sow Thistle	198
SPARGANIACEÆ	106
Sparganium	106
Spartina	115
Spear-grass	117
Specularia	197
Speedwells	193
Spergula	143
SPERMOPHYTES	15
Spiderwort	126
Spiesia	163
Spikenard	128
Spike-rushes	120
Spiræa	154
Spiranthes	131
Spirocela	126
Spleenwort	104
Sporobolus	112
Spruce	105
Spurey	147
Spurge Family	163
Spurges	167
Squaw Currant	153
Squirrel-tail-grass	119
Stachys	189
Staff-tree Family	169
Staghorn Sumac	169
Stanleya	148
Star-grass	130
Steironema	181
Stellaria	142
Stickseeds	186
Sticktight	209
Stipa	111
Stitch-wort	142
St. Johns-wort Family	171

	PAGE		PAGE
Stock	151	Tower Mustard	151
Stonecrop	152	Tradescantia	126
Strawberry	154	Tragopogon	198
Streptopus	129	Trailing Mahonia	147
Strong-scented Eragrostis	116	Trifolium	159
Strophostyles	165	Triglochin	108
Sugar Maple	169	Trillium	129
Sumac Family	169	*Troximon*	198
Sunflower	208	Tumble Weed	140
Sweet Cicely	178	Tumbling Mustard	148
Sweet Clover	159	Turkey-foot Grass	109
Sweet Coltsfoot	211	Twinflower	195
Sweet Flag	125	Twist-foot	129
Sweet Locust	158	Typha	106
Swertia	182	Typhaceæ	106
Symphoricarpus	196	ULMACEÆ	134
Syntherisma	109	Ulmus	134
Synthris	192	Umbrella-wort	140
Talinum	141	UMBELLIFERÆ	177
Tall Marsh-grass	115	Umbelworts	177
Tanacetum	210	Unifolium	129
Tansy	210	Urtica	135
Tansey Mustard	150	URTICACEÆ	135
Tape-grass	108	Urticastrum	135
Taraxacum	198	Utricularia	194
Tare	165	Uvularia	127
Tellima	153	Vaccaria	141
Teucrium	188	VACCINEACEÆ	180
Texas Thistle	191	Vaccinium	180
Thalaspi	148	Vagnera	128
Thalesia	194	Valerian	197
Thalictrum	147	Valeriana	196
Thermopsis	159	VALERINACEÆ	196
Thin-grass	113	Vallisneria	108
Thistle Family	200	VALLISNERIACEÆ	108
Thistles	212	Venice Mallow	171
Thorn apple	157	Venus-hair fein	104
Thoroughwort	201	Verbascum	191
Tickseed	109	Verbena	187
Tick-trefoils	164	VERBENACEÆ	187
Timothy	112	Vernonia	200
Tilia	171	Veronica	192
TILIACEÆ	171	Vervain Family	187
Toad Flax	191	Vervains	187
Toad Rush	126	*Vesicaria*	150
Tobacco Root	196	Vetch	164
Touch-me-nots	170	Vetchlings	165

	PAGE
Viburnum	195
Vicia	164
Viper's Bugloss	187
Viola	172
VIOLACEÆ	172
Voilet Family	172
Virginia creeper	170
Virginian Grape-fern	103
Virgin's Bower	145
Viscid Aster	205
VITACEÆ	170
Vitis	170
Wahoo	169
Wake-robin	129
Wall-flower	152
Walnut	131
Washingtonia	178
Water-cress	149
Water-crowfoot	146
Water Fennel	168
Water Helmock	179
Water Hemp	140
Water Hoarhound	190
Waterleaf Family	185
Water Lilies	143
Water Millfoil Family	176
Water-Plantain	108
Water Starwort Family	268
Water Star-grass	126
Waterwort	172
Water-whirl Grass	116
Western Blight	139
Western Ruppia	107
Western Stipa	111
Western Water Hemp	140
Western Wheat Grass	119
Western Yellow Pine	105
Wheat-grass	118
White Ash	181
White Campion	141
White Clover	159
White Daisy	210

	PAGE
White Elm	134
White Grass	110
White Mustard	148
White Sage	139
White Spruce	105
Whitlow-grass	150
Wild Bean	165
Wild Bergamot	189
Wild Carrot	177
Wild Leeks	127
Wild Licorice	164
Wild Oats	114
Wild Onions	127
Wild Parsnip	177
Wild Peas	165
Wild Plum	157
Wild Rice	110
Wild Rye	119
Wild Sarsaparilla	177
Willow Herb	174
Willows	132
Wind Flower	145
Wintergreen Family	180
Witch Grass	110
Wolfberry	196
Wolfsbane	144
Woodbine	170
Wood Nettle	135
Woodsia	103
Wood-sorrel Family	166
Wormwood	210
Wultenia	193
Xanthium	200
Xanthoxylum	167
Yarrow	210
Yellow Pine	105
Yucca	128
ZANNICHELLIA	107
Zizania	110
Zizia	178
Zygademus	127

ERRATA.

Page 101, 7th line, for herberium, read herbarium.
Page 103, 5th line, for matricarifolium, read matricariaefolium.
Page 106, 4th line, insert (L) after Virginiana.
Page 107, 30th line, for occideutalis, read occidentalis.
Page 110, 8th line, for Capillare, read capillare.
Page 110, 14th and 15th lines, for Viridis and *Viridis*, read viridis and *viridis*.
Page 111, 20th line, for Spartea, read spartea.
Page 112, 14th line, for Sylvatica, read sylvatica.
Page 114, 5th line, for macouniana, read Macouniana.
Page 114, 39th line, for Striata, read striata.
Page 115, 1st line, for Danthona, read Dauthonia.
Page 115, 3rd line, for Cynosuroides, read cynosuroides.
Page 115, 19th line, for Curtipendula, read curtipendula.
Page 116, 4th line, for purshii, read Purshii.
Page 116, 6th line, for Major, read major.
Page 118, 24th line, Cough-grass, read Couch-grass.
Page 119, 15th line, for Jubatum, read jubatum.
Page 119, 30th line, for Elymoides, read elymoides.
Page 126, 2nd line, for Spirodella, read Spirodela.
Page 126, 11th line, for virginiana, read Virginiana.
Page 128, 27th line, for Officinalis, read officinalis.
Page 128, 30th line, for Spiknard, read Spikenard.
Page 129, 6th line, for amqlexicaulis, read amplexicaulis.
Page 129, 22nd line, for Commutatum, read commutatum.
Page 130, 12th line, for augustifolium, read angustifolium.
Page 130, 27th, 31st, 34th lines, for Cyprepedium, read Cypripedium.
Page 133, 17th line, for virginica, read Virginica.
Page 139, 7th and 8th lines, for Cycloma, read Cycloloma.
Page 141, 16th line, for Silena, read Silene.
Page 141, 19th and 21st lines, Lychinis, read Lychnis.
Page 141, 19th line, for Champion, read Campion.
Page 142, 12th, 14th, 17th and 20th lines, for Silena, read Silene.
Page 142, 12th, 14th and 17th lines, for Champion, read Campion.
Page 143, 24th line, for oderata, read odorata.
Page 144, 28th line, for Fisher's, read Fischer's.
Page 145, 5th line, for virginiana, read Virginiana.
Page 150, 15th line, for Shepard's, read Sheperd's.
Page 153, 12th line, for Setosum, read setosum.
Page 153, 15th line, for Kiber, read Ribes.
Page 154, 12th line, for *nutkanus*, read *Nutkanus*.
Page 154, 31st line, for *Vesca*, read *vesca*.
Page 155, 15th line, for hippiana, read Hippiana.
Page 155, 20th line, for fruiticosa, read fruticosa.
Page 156, 9th line, for Macrophyllum, read macrophyllum.
Page 156, 17th line, for *Agrimonia*, read *Agromonia*.
Page 156, 20th line, for Agrimonia, read Agromonia.
Page 156, 28th line, for woodesii, read Woodsii.
Page 161, 17th, 21st, 26th and 31st lines, for *Petalstemon*, read *Petalostemon*.
Page 165, 21st and 22nd lines, for Apols, read Apios.
Page 168, 19th line, for dictyosporma, read dictyosperma.

Page 168, 25th line, for *mountain*, read *montana*.
Page 169, 16th line, for atropurpureous, read atropurpureus.
Page 170, 27th line, for Vulpina, read vulpina.
Page 171, 4th line, for Tilliaceœ, read Tiliaceæ.
Page 171, 10th line, for Sylvestris, read sylvestris.
Page 177, 5th line, for carrota, read carota.
Page 178, 1st line, for canadensis, read Canadensis.
Page 178, 2nd line, for *canadensis*, read *Canadensis*.
Page 179, 13th line, for canàdensis, read Canadensis.
Page 179, 18th line, for *augustifolia*, read *angustifolia*.
Page 180, 8th line, for Elliptica, read elliptica.
Page 184, 20th line, for Epithymum, read epithymum.
Page 185, 29th line, for Curassavicum, read curassavicum.
Page 187, 24th line, for Molle, read molle, for Cromwell, Gromwell
Page 189, 28th and 30th lines, for Bergamont, read Bergamot.
Page 203, 25th line, for Radula, read radula.
Page 209, 17th line, Cereopsis, read Coreopsis.

www.ingramcontent.com/pod-product-compliance
Lightning Source LLC
Chambersburg PA
CBHW021937160426
43195CB00011B/1128